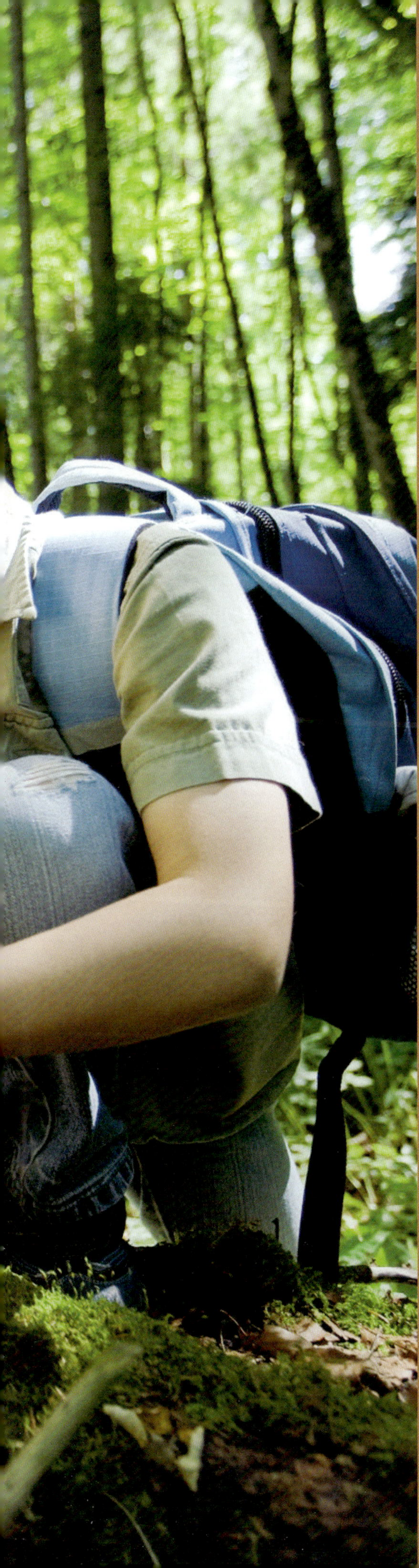

zero creatives GmbH/Photo Researchers

# 1
# Chapter

# Introduction to Science

Equipped with his five senses, man explores the universe around him and calls the adventure Science.

— Edwin Hubble

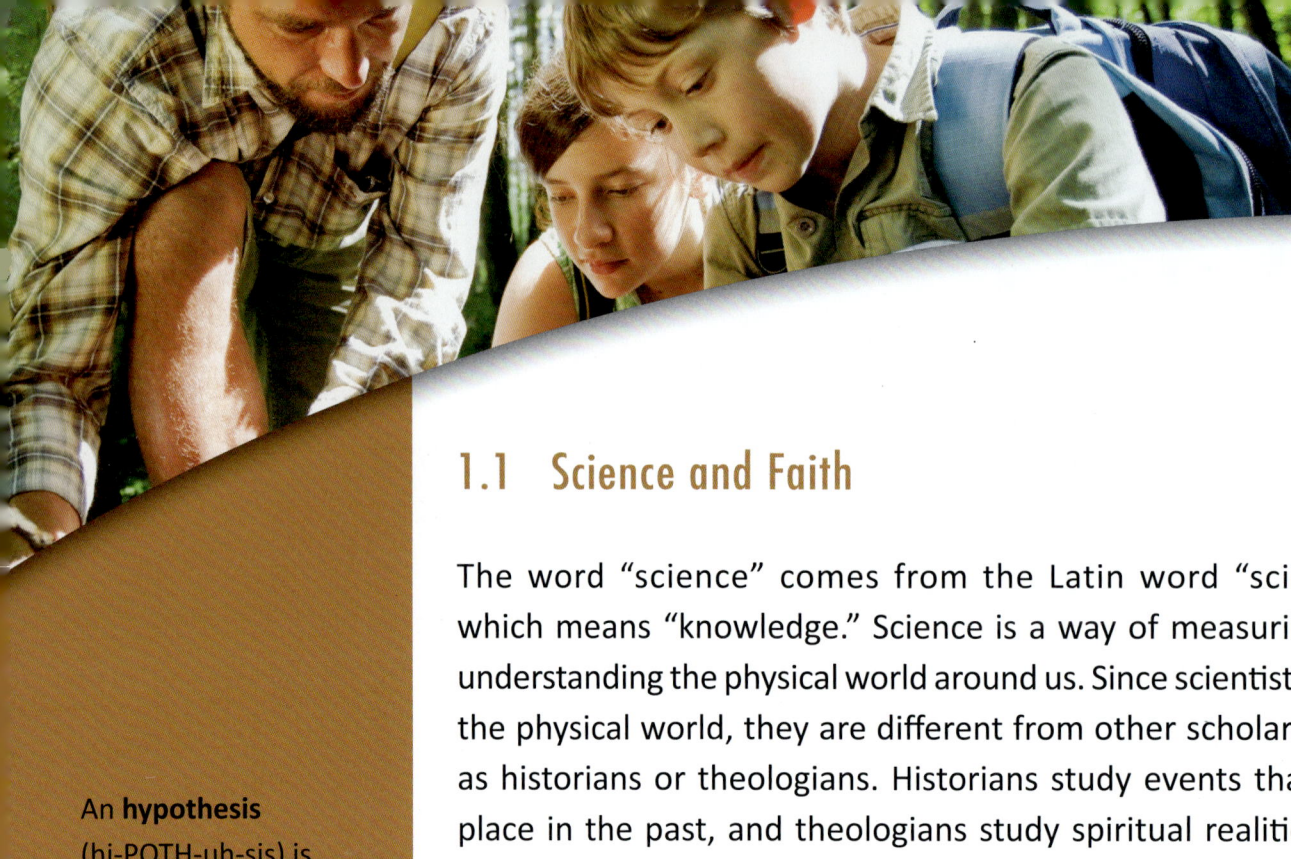

## 1.1 Science and Faith

The word "science" comes from the Latin word "scientia," which means "knowledge." Science is a way of measuring and understanding the physical world around us. Since scientists study the physical world, they are different from other scholars, such as historians or theologians. Historians study events that took place in the past, and theologians study spiritual realities, but scientists study things that can be seen, touched, tasted, smelled, and heard. When a scientist has a question about the physical world, he forms an **hypothesis**, and then uses experiments to test whether his hypothesis is correct.

You may wonder what a scientist does when he has a question that cannot be answered through physical experiments. For instance, what does the scientist do if he wants to know whether angels exist? Since angels do not have bodies, they cannot be seen, touched, tasted, smelled, or heard. This means that there is no experiment a scientist can perform to answer the question of whether or not they exist. So what is he to do?

Well, what would you do if you were writing a book report on Mark Twain's *The Prince and the Pauper,* and you needed to know the dates when Henry VIII was king of England? You would turn to your history book and look up the information you needed. You wouldn't complain because your literature text wasn't also a history text. On the contrary! You would be grateful that you had access to both a literature text and a history text. Without both resources, you would not be able to fully enjoy Mark Twain's book.

In the same way, a scientist who is wise does not complain when

An **hypothesis** (hi-POTH-uh-sis) is a scientific guess.

A CATHOLIC AND HANDS-ON APPROACH TO SCIENCE

# BEHOLD AND SEE 6

## RoseMary C. Johnson

## About the Author

**RoseMary Johnson** was homeschooled from kindergarten through high school and graduated *summa cum laude* from the University of Dallas, where she is currently pursuing a Ph.D. Since she has always been interested in science, she is grateful for the opportunity to author *Behold and See 6* and to share her love of science with other homeschoolers. She is also the author of *The Secret Code of Poetry,* an introduction to literature for 7th–9th graders.

## Consultants

**Nancy Nicholson** is the inspiration behind the *Behold and See* science series and the author of *Behold and See 1* and *Behold and See 2*. Equipped with an abundance of God-given talent, a major in Secondary Education–English, and years of experience homeschooling her own children, she has written over thirty "For Little Folks" educational titles that have enriched thousands of students.

**Mary Reidy** is a homeschool alumna who credits her parents with first inspiring her love for science. She received a B.S. in Physics from the University of Dallas and a second B.S. in Mechanical Engineering at the University of Missouri–St. Louis. She has conducted research in nonlinear fluid dynamics, and she currently works in designing and manufacturing laboratory diagnostic equipment.

**Amanda Beheler** received her undergraduate and masters degrees in biology from the College of William and Mary where she worked with the Center of Conservation Biology on a variety of avian research projects. She received her Ph.D. in Wildlife Ecology from Purdue University. Currently, she is pursuing her greatest joy of raising and homeschooling her children.

**Richard Olenick** has taught physics and astronomy at the University of Dallas for over thirty years. During the 1980s, he was Associate Project Director of the PBS television series "The Mechanical Universe" and "Beyond the Mechanical Universe" and was the principle author of the accompanying textbooks. His research interests include cataclysmic variable research, modeling non-linear phenomena, and physics education.

## Credits

Illustrations and diagrams by AnneMarie Johnson

"St. Albert the Great" by Anne Simoneau (pg. 3)

Image credits may be found on page 355.

ISBN: 978-0-9851642-1-8

© 2012 Catholic Heritage Curricula

This book is under copyright. All rights reserved. No part of this book may be reproduced in any form by any means–electronic, mechanical, or graphic–without prior written permission. Thank you for honoring copyright law.

For more information:
　Catholic Heritage Curricula
　www.chcweb.com

Printed by Transcontinental Interglobe
Beauceville, Quebec, Canada
September 2020

For detailed, daily lesson plans, see *CHC Lesson Plans for Sixth Grade* and *Behold and See 6 Daily Lesson Plans* at www.chcweb.com.

# Table of Contents

Chapter 1: Introduction to Science, *1*

**Unit 1: Matter and Energy**

Chapter 2: Mass, Density, and the Classification of Matter, *7*

Chapter 3: Change of State and Solubility, *23*

Chapter 4: Types of Energy, *43*

Chapter 5: Electricity and Magnetism, *61*

Chapter 6: Simple Machines, *87*

**Unit 2: Biomes**

Chapter 7: Introduction to Biomes, *107*

Chapter 8: Arctic Tundra, *121*

Chapter 9: Boreal Forest, *145*

Chapter 10: Forests: Temperate and Tropical, *169*

Chapter 11: Grasslands and Deserts, *191*

Chapter 12: Stewards of Creation, *217*

**Unit 3: Astronomy**

Chapter 13: Constellations and Star-Gazing, *233*

Chapter 14: Gravity, Orbits, and the Moon, *261*

Chapter 15: The Sun and the Other Stars, *285*

Chapter 16: The Solar System, *307*

Chapter 17: Galaxies and the Universe, *333*

The accompanying student workbook provides student-friendly exercises, research assignments, experiments, and tests.

he encounters a question that cannot be answered by physical experiments. Instead, he is grateful for the other sources of truth that God has given him, especially for Divine Revelation. What scientists are not able to know about God and about human nature through science, they are able to know through faith. This is why St. John Paul II described faith and reason—religion and science—as "two wings on which the human spirit rises to the contemplation of truth." No wonder many of the greatest scientists have also been devout Christians!

## 1.2 About the Text

Did you know that you can lift an elephant with one hand if you use enough moveable pulleys? Did you know that there are plants that look like pebbles and insects that look like dead leaves? Did you know that the gold in your parents' wedding rings was forged inside an explosion 700 septillion (700,000,000,000,000,000,000,000,000) times more powerful than the most powerful nuclear weapon? You will, once you have completed Units 1, 2, and 3 of *Behold and See 6*!

*Behold and See 6* focuses on physical science, ecology, and astronomy. Physical science includes chemistry and physics; in Unit 1 you will be studying and doing experiments with matter and energy, electricity and magnetism, levers and pulleys, and similar topics.

In Unit 2 you will tour the biomes of the world, meeting creatures as amusing as the shovel-snouted lizard and as sinister as the strangler fig. Along the way, you will learn about interdependence, food chains, climates, and the remarkable "equipment" God has given His creatures to help them survive.

### St. Albert the Great

**Famous Catholic Scientists**

**St. Albert the Great** (1206–1280) is the patron of the natural sciences and contributed to many scientific fields, including chemistry, biology, astronomy, and meteorology.

**Francesco Maria Grimaldi** (1618–1663) was a Jesuit mathematician and astronomer. The Grimaldi crater on the Moon is named after him because he drew one of the first accurate maps of the Moon's surface.

**Bl. Niels Stensen** (1638–1686) made major breakthroughs in both anatomy and geology. He was also a very holy bishop—so holy, in fact, that he was beatified in 1987.

**Jerome Lejeune** (1926–1994) discovered the genetic cause for Down Syndrome and zealously defended the sanctity of all human life. The cause for his canonization was opened in 2007.

**John A. O'Keefe** (1916–2000), a devout Catholic, was a major lunar space scientist and a leader in NASA's space program.

**Fr. Stanley Jaki** (1924–2009), a Benedictine priest, was a brilliant thinker with doctoral degrees in both theology and physics.

In the third and final unit, you will discover the wonders of astronomy, and use what you have learned in Units 1 and 2 to explore the night sky, the solar system, the galaxy, and the universe. Get your star charts out and prepare for some late-night star-gazing!

The *Behold and See 6* Student Text is designed to work hand in hand with the experiments and workbook pages in the *Behold and See 6* Workbook. Small symbols have been placed throughout the chapters to indicate when it is time to do an experiment or complete a workbook assignment. It is crucial, especially for the first and third units, that you alternate reading the text with doing experiments and completing workbook pages. This step-by-step approach will allow you to develop a real understanding of the topics presented, instead of just reading scientific facts.

The "Check It Out!" web links listed throughout the workbook may be explored as time allows. Interactive links, extension activities, and video clips can make difficult concepts easier to grasp.

## 1.3  Science Notebook

Before you dive into Unit 1, take a few days to set up your Science Notebook. Your Science Notebook is the place where you will write your notes and record the results of your experiments.

On the title page of your notebook, write the name of the course, your name, where you live, and the month and year. You may wish to cut out photos from magazines and catalogs to decorate this page, or draw your own pictures. Ideas include pictures of magnets, simple machines, atoms, and electricity; plants and animals; the Sun, stars, and Moon.

On the next page, write "Unit 1: Matter and Energy." You can decorate this page later, once you have completed Unit 1 and have a better understanding of what the unit includes. Later in the year you can create similar title pages for Units 2 and 3.

The rest of the pages are for you! They provide a place for you to record the results of your experiments and complete assignments given in the workbook. Whenever you add something to your Science Notebook, make sure you add a title that explains what you have added. For instance, in Experiment #1 you are asked to record the weights of 10 different ingredients. If you simply listed the numbers in your notebook without labeling them, you would quickly forget what the numbers were for! To prevent this, always begin by writing the name of the experiment or assignment at the top of the page, and be sure to provide labels for any information, charts, and drawings.

## 1.4  Final Notes

The experiments and activities that begin on page 117 in the workbook are a large part of this course, so it is important that you do as many of them as you can. The experiments will proceed more smoothly if you read through the instructions before beginning. You wouldn't want to start an experiment on a rainy day and discover, halfway through, that the experiment has to be completed outside.

The experiments have been carefully selected to ensure that they can all be done using household materials. It is a good idea to check the supply list on pages 117-118 of the workbook so that you can start saving bottles, corks, and other supplies.

Finally, make sure you get parental permission before you do any experiments that require the use of a sharp knife, a stove, or other, potentially dangerous materials. Experiments that require parental supervision have been noted as such in the workbook.

UNIT 1

# 2
Chapter

# Mass, Density, and the Classification of Matter

Scientists do not create the world; they learn about it and attempt to imitate it, following the laws and intelligibility that nature manifests to us.

— Pope Benedict XVI

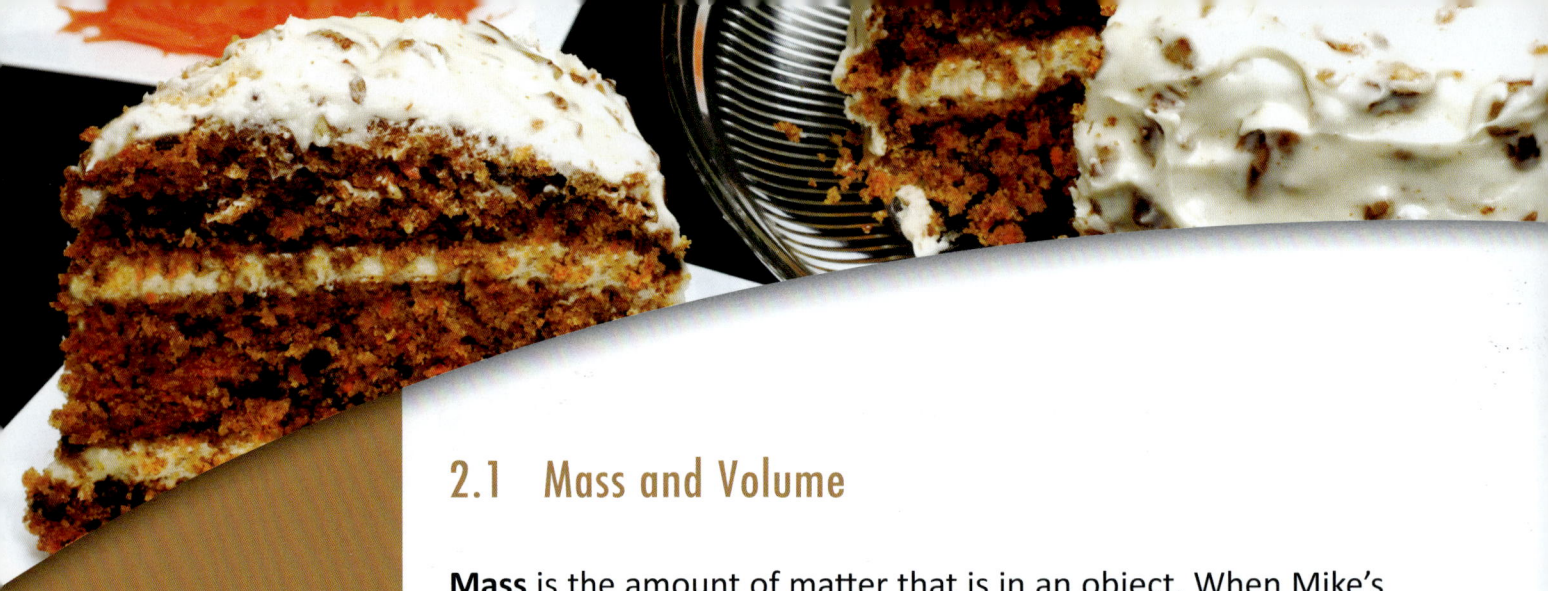

## 2.1 Mass and Volume

**Mass** is the amount of matter that is in an object. When Mike's mother adds extra nuts to his favorite Nutty Carrot Cake, the cake becomes more massive—that is, it has more matter, more "stuff," in it.

The cake also has a greater **volume** after Mike's mother adds the extra nuts. This means that it takes up more space. The volume of an object is how much space it takes up.

Mike used to think that the volume of an object depended on its mass. After all, when his mother added more nuts, the cake batter took up more space. So if an object is very massive, he thought, it will be very big. If it has only a little mass, it will be small.

But then Mom told him to watch the cake after she put it in the oven. As Mike watched the cake through the oven window, the batter slowly rose. By the time the kitchen timer beeped and Mom carefully took the cake out of the oven, the batter had risen far above the edge of the pan.

"See how much bigger the cake is now that it is baked," Mom remarked.

Mike was confused. He had watched his mother bake before, so he was not surprised that the cake had expanded in the oven. But he had never thought about what happened to the cake's mass and volume as it expanded. "How could the cake's volume increase when we didn't add any more matter?" Mike asked.

---

**Mass** is the amount of matter that is in an object.

The **volume** of an object is how much space it takes up.

Mike's mother smiled. "If you can't figure it out, you'll have to wait until tonight when we eat the cake."

After dinner that evening, Mike's father passed him a thick slice of Nutty Carrot Cake. "After you've eaten a few bites to energize your brain, examine the inside of the cake carefully. This cake is evidence that the volume of an object can change even if its mass does not."

Mike ate a few bites of cake—mmm, the crunchy nuts contrasted perfectly with the soft, fluffy cake. "Eureka!" Mike exclaimed. "The cake has a larger size because it is so fluffy. Even though we didn't add any extra matter, the original matter is taking up more room because it is full of air bubbles!"

"Before Mom baked the cake, it was short and flat," Mike explained to Nick and Christie, his younger siblings. "When Mom baked it, lots of air bubbles were formed in it and it grew larger.

"Now, if I smash the cake with my fork," Mike continued, "the cake gets small again because I've made all the air bubbles disappear. But even though it takes up less room, there is still just as much cake on my plate.

"This shows that the volume of an object does not depend on its mass," Mike concluded. "The volume of an object can change even when its mass stays the same."[1]

"Great job," his mother praised him. "You're thinking like a real scientist."

*Eureka* (yuh-REE-kuh) is a Greek word meaning "I have found it." The story goes that Archimedes, a famous Greek scientist from the 3rd century B.C., shouted *Eureka* when he made a breakthrough in understanding buoyancy and density. You will learn about these topics later in this chapter.

---
[1] Air bubbles have a little bit of mass, but not enough to make a difference for this experiment. We'll learn more about this in Chapter 3.

"Yes," Mike's father agreed. "In fact, you went through all the steps of the scientific method.

"First you **observed** the cake expanding in the oven.

"Then you formed a **question**. You asked why the cake got bigger when we hadn't added any matter.

"When you saw the fluffy texture of the cake, you guessed that the cake grew in size because air bubbles made its mass fill a larger space. This was your **hypothesis**.

"Then you came up with a **method** to test whether your hypothesis was correct. By pressing on the cake with your fork, you tested what would happen if you flattened all the air bubbles. Sure enough, the cake became small again.

"Next, you explained the results of your experiment to Nick and Christie. In other words, you **interpreted** the results of your experiment.

"Finally, you drew a **conclusion**: the volume of an object does not depend on its mass."

"Wow!" Mike exclaimed. "The scientific method seemed so complicated when you were teaching me about it yesterday, but it's pretty much what I do already when I'm wondering about something."

**The Scientific Method**
1. Observation
2. Question
3. Hypothesis
4. Method
5. Interpretation
6. Conclusion

"Yes, indeed," his father agreed. "Put simply, science is a **systematic** way of exploring the world that God has created. This is something that everyone can do who knows how to wonder at the beauty of the world around him."

"But science is only possible because God created the world according to an orderly design, and gave us our minds to understand His design," Mom added. "I've sometimes thought that there should be a seventh step in the scientific method: thanksgiving for an orderly, beautiful world and for our ability to understand it."

"Absolutely!" Mike agreed. Scraping up the last bit of cream cheese frosting from his plate, he added, "And in the case of this experiment, we should thank the cook, too. That was the best cake I've ever tasted, Mom!"

## 2.2 Density

Since the volume of an object does not depend on its mass, an object can have a lot of matter in it, but still be quite small (for instance, a brick). Or it can contain only a little matter, but still be quite large (for instance, a pillow).

The measurement of how much matter is packed into a certain space is called its **density**. When an object contains a lot of matter for its size, we say that it is very dense, or has great density. When an object contains only a little matter for its size, we say it is not dense, or has little density.

**Systematic** means "step by step, in an organized way."

**Density** is the measurement of how much matter is packed into a certain space.

 **Workbook pg. 1**

## 2.3 Gravity and the Buoyant Force

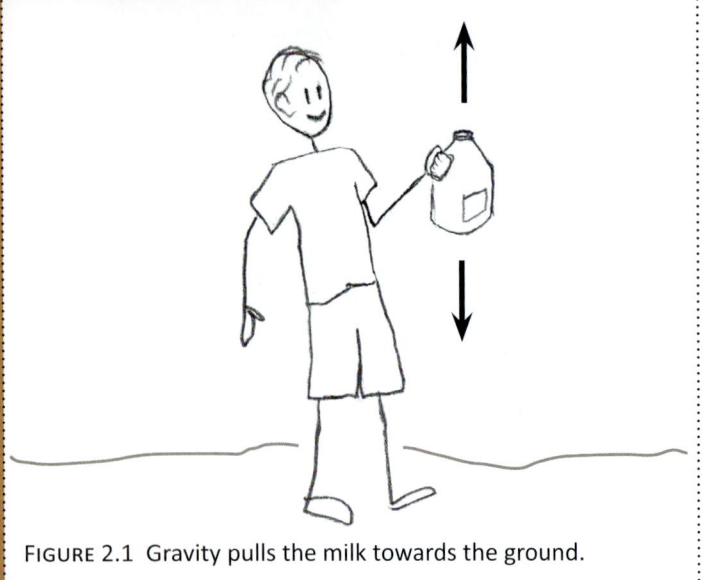

FIGURE 2.1 Gravity pulls the milk towards the ground.

"Mom, why does oil float when it is mixed with water?" Mike asked.

"To answer that question we'll need to talk about gravity," Mom replied. "To begin, let's think about how a gallon of milk feels when you're helping to unload the groceries. It feels heavy, right?"

"Boy, does it!" Mike exclaimed. "Last week I tried to carry two in one hand, and I almost dropped them. I had to set them down and take one in each hand."

"Why did they feel that way?" Mom asked.

"Because the Earth's gravity was pulling the milk down to the ground harder than I could pull it up from the ground. It was like we were playing tug-of-war!" Mike explained, imagining a David-and-Goliath battle between himself and the Earth.

"Excellent! But can you explain why the bag of tortilla chips felt so much lighter than the gallon of milk? Wasn't gravity pulling on it, too? Or was the Earth only hungry for milk, and not chips?" Mom teased.

"That's a tough question," Mike admitted. "A bag of chips is mostly air, though. Maybe it's because a gallon of milk has more mass than a bag of chips."

"You've hit the nail right on the head!" Mom praised. "Mass is the amount of matter in an object, and **gravity** is a force that acts

**Gravity** is a force that acts on matter.

on matter. It doesn't act on empty space.

"So the more matter there is in an object, the greater the force of gravity will be for the object. Since a gallon of milk has more mass than a bag of chips, there is more of it for gravity to 'pull' on. This explains why it feels so heavy to us."

"That makes sense," Mike agreed.

"Now, imagine a cupful of water and a cupful of oil. Can you guess which weighs more?"

Mike closed his eyes and thought for a few seconds. "Oil," he guessed. "It's thicker, so it has more matter."

"That's a great hypothesis. Let's test it!" Mom took out her kitchen scale, a measuring cup, and a bottle of vegetable oil.

Mike filled the measuring cup with water, weighed it on the scale, and wrote its weight down on a piece of paper. Then he filled the measuring cup with oil and set it on the kitchen scale. "Mom," he exclaimed, "the oil weighs less! The cup and the water together weigh 11 ounces, but the cup with oil weighs only 10½ ounces. The water is about half an ounce heavier than the oil!"

"Good job! This means that gravity is pulling harder on the water than it is on the oil. Since gravity is a force that acts on matter, one cup of water must have more matter in it than one cup of oil does. In other words, water must be denser than oil.

"This is why oil floats on water. Both water and oil are pulled towards the ground by gravity, but the water is pulled harder, because it is denser. As the dense water is pulled to the ground by gravity, it 'pushes' the oil out of the way, up towards the surface.

The upward "push" which water (or any liquid) gives to objects is called the **buoyant** (BOY-uhnt) **force**.

When an object is denser than water, it sinks. When an object is less dense, it floats.

To **displace** means to push out of the way.

"The 'push' which water gives to objects is called the **buoyant force**. When an object is less dense than water, the buoyant force makes it float. This is what happens with oil.

"When an object is denser than water, the buoyant force is still present, but it is not strong enough to keep the object from sinking. Instead, the *object* pushes the *water* out of the way. This is what happens when you drop a stone into a pond."

"But why do huge iron ships float on the ocean?" Mike asked, puzzled. "Iron is much heavier than water."

"Good question! Solid iron is certainly denser and heavier than water. Iron ships can float because they are shaped in a way that makes the boat as a whole less dense than water.

"Try this: get out a ball of clay and drop it in a bowl of water. It sinks, right? That's because it's heavier than the water that it **displaces**, or pushes out of the way.

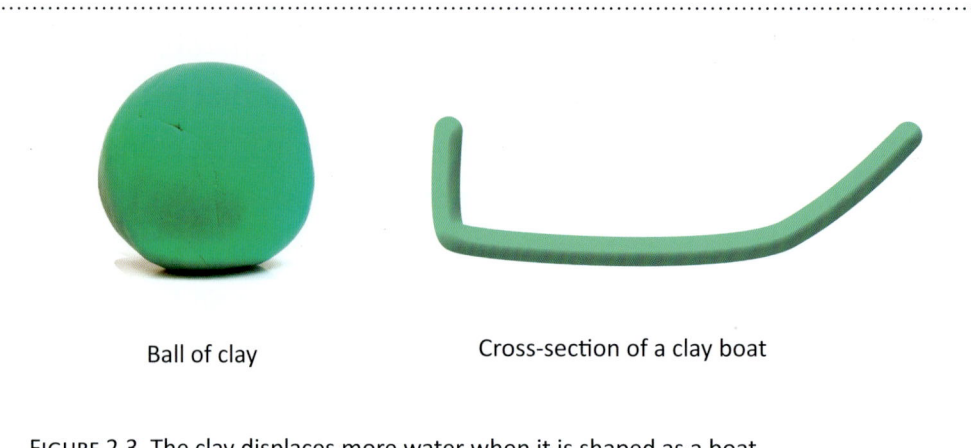

Ball of clay          Cross-section of a clay boat

FIGURE 2.3  The clay displaces more water when it is shaped as a boat.

14

"Now mold the clay into an empty boat. In this shape, the clay floats. You can even fill it with a few coins or pebbles without making it sink.

"The reason this happens is that the clay boat is hollow, so its overall density is quite low. Density, you remember, is the measurement of how much matter is packed into a certain space. When the clay is shaped like a boat, its density as a whole (including its hollow interior) is less than the density of the water it displaces. This means there is more matter in the displaced water than in the clay boat, so the force of gravity acts more strongly on the displaced water and the water pushes the boat upwards through the buoyant force."

When Mike gives Christie a "porpoise-ride" in the swimming pool, she feels much lighter than she does when he gives her a "horsie-ride" in the park. This is because of the buoyant force. In the park, Mike has to carry all of Christie's weight by himself. In the water, the buoyant force does some of the work for him by "pushing" her up towards the surface.

 Workbook pgs. 2-3

 Experiments #1-3 (workbook pgs. 119-120)

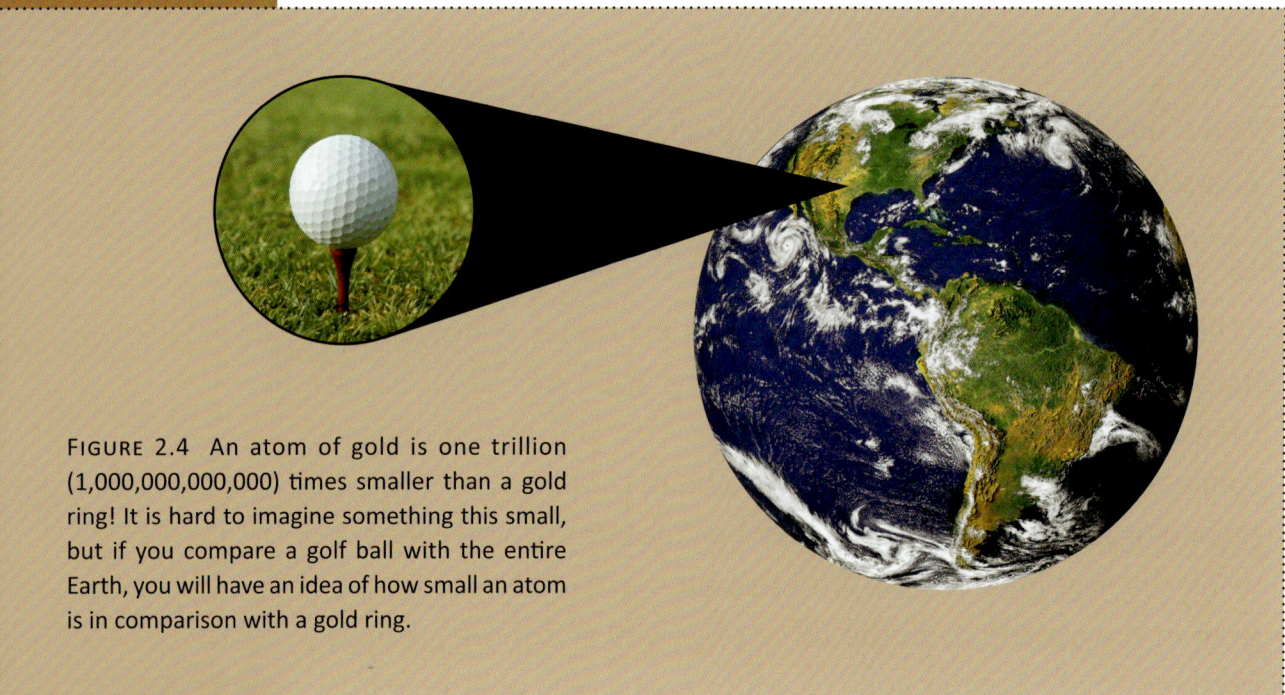

FIGURE 2.4 An atom of gold is one trillion (1,000,000,000,000) times smaller than a gold ring! It is hard to imagine something this small, but if you compare a golf ball with the entire Earth, you will have an idea of how small an atom is in comparison with a gold ring.

An **atom** is a tiny particle and the basic building block of matter.

An **element** is made of only one type of atom.

## 2.4 Elements, Compounds, and Mixtures

All matter is made out of tiny particles called **atoms**. Atoms are so small that they can only be seen with the most powerful microscopes. Atoms are the basic building blocks of matter. Objects have the properties that they do because of how their atoms are arranged.

For instance, when Mike's older sister Daniella fixes her siblings a pitcher of ice-cold lemonade, she combines lemon juice, sugar, and water. All of these ingredients are made out of atoms, but each ingredient is very different from the others. Water is not the same as sugar. Sugar is not the same as lemon juice. Lemon juice is not the same as water. The reason that sugar, water, and lemon juice look and taste different from each other is that they are made out of different combinations of atoms.

There are three main ways in which atoms can be combined. First, you can combine millions of the same type of atom to form an **element**. Pure gold is an element because it is made out of millions of atoms of gold and nothing else. Oxygen is an element because it is made of nothing but oxygen atoms, and carbon is an element because it is made of nothing but carbon atoms. Since there are only about 100 different kinds of atoms in the whole world, there are only about 100 different elements in the world, too.

Atoms can also be combined to form **compounds**. A compound contains at least two different kinds of atoms. For instance, water is a compound that is made out of oxygen atoms and hydrogen atoms. The atoms in a compound aren't mixed together randomly, though. The different types of atoms in a compound are chemically joined together into **molecules**. Molecules are tiny particles that are formed when two or more different atoms are chemically "glued" together in a particular position. For example, a molecule of water is made of one oxygen atom and two hydrogen atoms chemically "glued" in the correct positions. If a molecule of water loses one of its atoms, it ceases to be a molecule of water and becomes something else, like a molecule of hydroxide.

Sucrose, or table sugar, is another, more complicated compound. A single sucrose molecule is made of 22 hydrogen atoms, 12 carbon atoms, and 11 oxygen atoms. When these atoms are linked together in exactly the right way, the result is a molecule of sucrose, or table sugar.

Finally, atoms can be combined to form **mixtures**. Mixtures are different from elements and compounds because they are not made out of only one type of particle. Elements can only contain one type of atom and compounds can only contain one type of molecule, but mixtures can contain any number of different atoms and molecules, in any order. In a mixture, the different particles are mixed together, but they are not chemically glued together in the same way as atoms and molecules. The atoms

A **compound** is made of only one type of molecule.

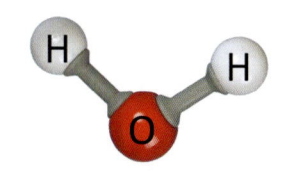

FIGURE 2.5 Model of a water molecule

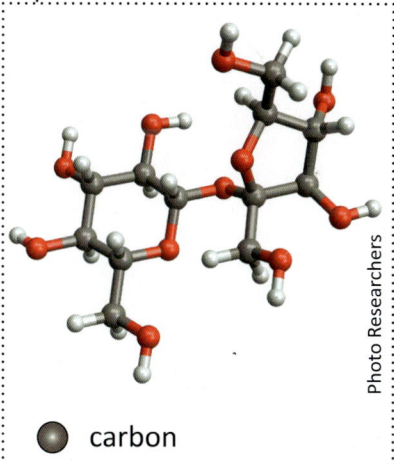

- carbon
- hydrogen
- oxygen

FIGURE 2.6 Model of a sucrose molecule

**Molecules** (MOL-uh-kyoolz) are tiny particles formed when two or more different atoms are chemically "glued" together in a particular position.

A **mixture** is made of any number of different atoms and molecules arranged in any order.

and molecules in a mixture don't link up into new types of particles.

Nutty Carrot Cake is a good example. When Mike's mom puts the different ingredients into the bowl and blends them together, they combine to form cake batter. But the atoms and molecules in the different ingredients don't join to form "cake-batter molecules." If we had microscopic vision, we could still see molecules of sugar, salt, water, and more in the Nutty Carrot Cake mixture.

The air we breathe is also a mixture. The air is made of several different gases, mostly nitrogen and oxygen. The atoms and molecules of these different gases swirl around each other, but they don't combine into new molecules.

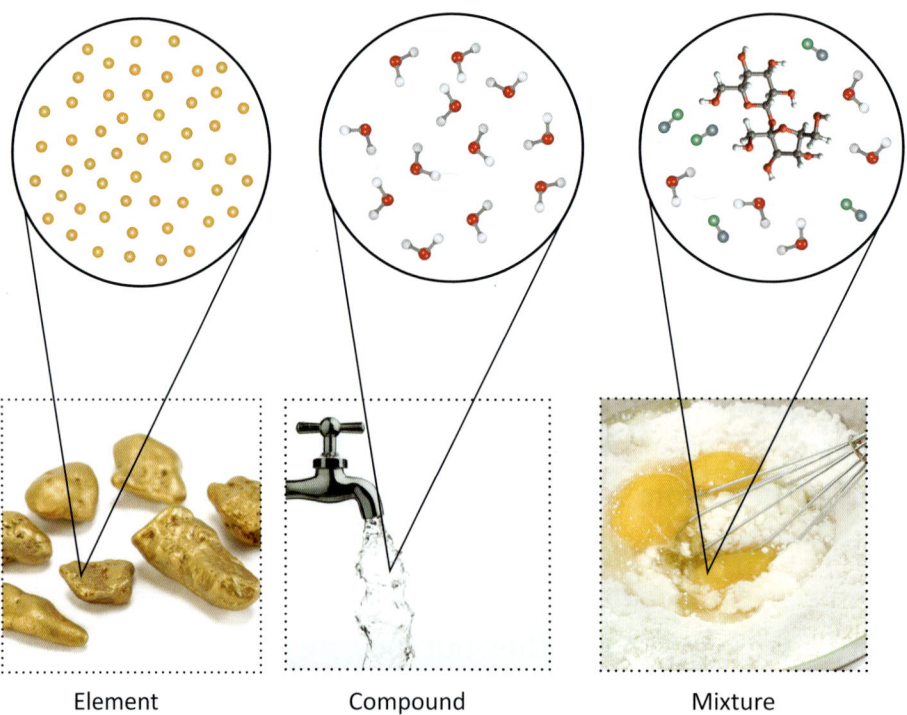

FIGURE 2.7  The molecules and atoms in gold, water, and cake batter

Elements and compounds are often called **pure substances**, because no matter what part of an element or compound you look at, you will always find the same atoms or molecules. A mixture is different from a pure substance because in a mixture you can find many different atoms and molecules, all mixed together.

A mixture is also unique because the different compounds and elements in the mixture can be separated from each other by purely physical processes. Physical processes include techniques such as sifting, filtering, and evaporation. The different substances in a compound, on the other hand, can only be separated from each other by chemical reactions that break up the molecules into smaller molecules or atoms.

A **pure substance** is matter that consists of only one type of particle. If a pure substance is made of atoms, it is an element; if a pure substance is made of molecules, it is a compound.

**Physical processes** are techniques that can be used to separate the different parts of a mixture. Physical processes do not involve chemical reactions.

**Chemical processes** are chemical reactions that can be used to separate a compound into its different parts.

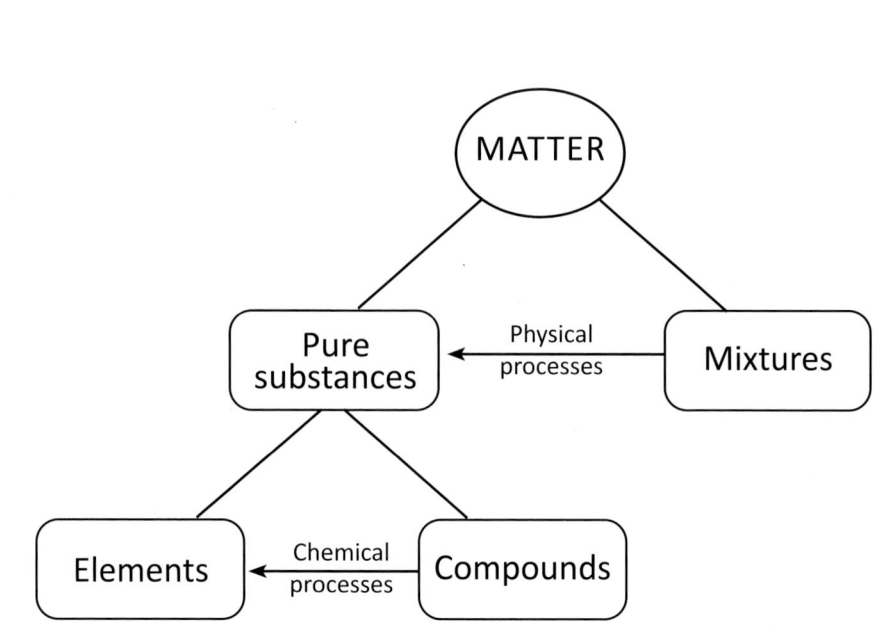

FIGURE 2.8 All matter is either a pure substance or a mixture, and every pure substance is either an element or a compound. Mixtures can be separated into pure substances through physical processes; compounds can be separated into elements through chemical processes.

 Workbook pgs. 4-5

 Experiments #4-5

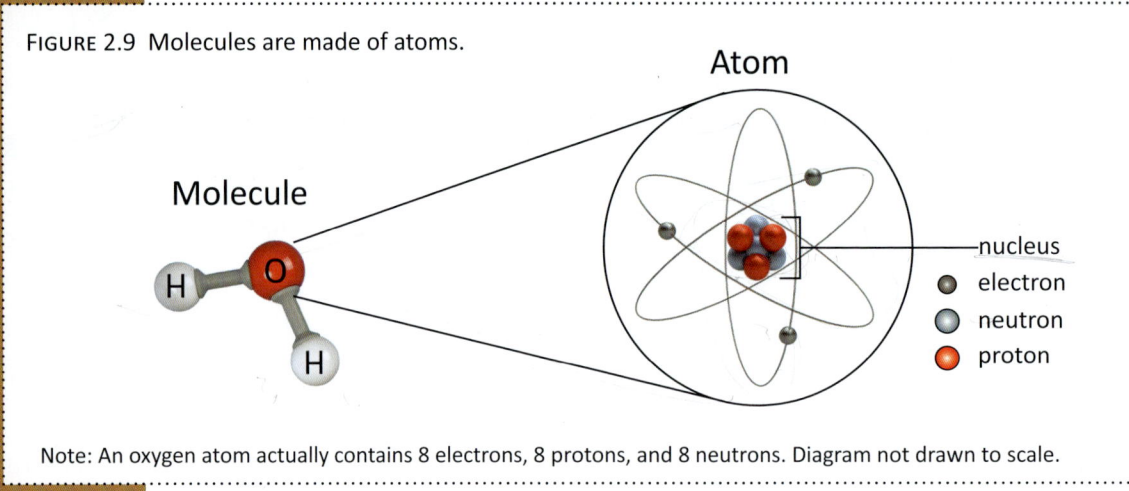

FIGURE 2.9 Molecules are made of atoms.

Note: An oxygen atom actually contains 8 electrons, 8 protons, and 8 neutrons. Diagram not drawn to scale.

## 2.5 Protons, Electrons, and Neutrons

"As you have learned, a mixture can be separated into its parts by physical processes, and a compound can be separated into its parts by chemical processes. But an element is made of only one type of atom, so to separate it into parts you would have to break apart the atom."

"Does an atom have parts?" Mike asked. "You said the atom was the basic building block of matter, so I thought it was the smallest thing that could exist."

"Well, atoms are pretty tiny," Dad agreed, "and atoms are the smallest possible pieces of a substance that can exist. But that's because if we break apart an atom of oxygen or an atom of gold, we don't have oxygen or gold anymore," Dad explained. "Instead we have a bunch of protons and electrons and neutrons."

"Proteins and electro-*whatrons*?" Mike gasped.

Dad chuckled. "Protons and electrons and neutrons. **Protons** and **neutrons** are tiny particles that form the center, or **nucleus**, of an atom. **Electrons** are even smaller particles that are constantly whizzing around the protons and neutrons."

"Wow! That sounds like a little solar system," Mike joked.

"Actually, that's not a bad way to think about an atom. Do you remember learning how far the Earth is from the Sun? Millions

---

**Protons** and **neutrons** (NOO-tronz) are tiny particles that form the center, or nucleus, of an atom. **Electrons** are even smaller particles that are constantly whizzing around the protons and neutrons.

The **nucleus** of an atom is its center, which is made up of protons and neutrons.

and millions of miles, right? And the distance from the Sun to the outer edge of the solar system is even greater. So most of the solar system is empty space.

"An atom is the same way—most of it is empty space. In fact, if there were an atom the size of a football stadium, the nucleus in the center would be about the size of a peanut. The electrons would be like mosquitoes whizzing around beyond the last row of seats. Everything in between would be empty space."

Mike's eyes were bugging out. "Mosquitoes? Peanuts? In a football stadium? And I thought atoms were small!"

"They are," Dad smiled. "They're just made of even smaller things. Isn't it amazing that God can make things that are so small, our eyes cross when we try to imagine them? And at the same time, He makes things like galaxies, which are mind-boggling-ly huge!"

"I guess that's because He's **infinite**!" Mike concluded. "Size doesn't make any difference to Him."

**Infinite** means not limited by a certain size or space.

Workbook pgs. 6-8

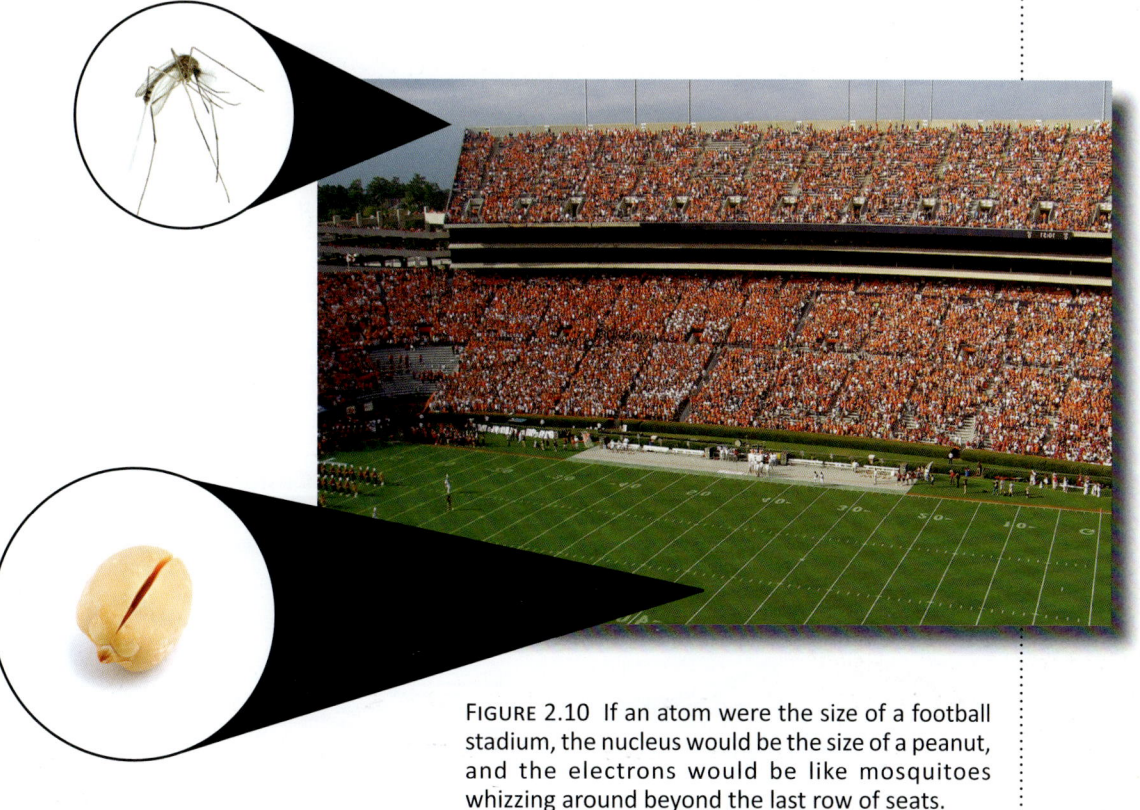

FIGURE 2.10 If an atom were the size of a football stadium, the nucleus would be the size of a peanut, and the electrons would be like mosquitoes whizzing around beyond the last row of seats.

# Chapter 3

# Change of State and Solubility

I am among those who think that science has great beauty. A scientist in his laboratory is not only a technician: he is also a child placed before natural phenomena which impress him like a fairy tale.
— Marie Curie

## 3.1 Energetic Atoms and Molecules

"Why do you think we are learning about tiny things like atoms?" Dad asked. "They're so small, they must be unimportant, right?"

"Of course not. They're the basic building blocks of matter!" Mike chanted.

"And these basic building blocks have a lot to do with how matter acts. Freezing, boiling, melting, evaporation, and all the other changes that matter goes through depend on the behavior of atoms and molecules. Right now we're going to study the **three states of matter**. Do you remember what they are?"

"The solid state, the liquid state, and the gaseous state," Mike listed.

"Exactly," Dad affirmed. "For instance, water can be ice, liquid water, or steam. And what usually makes matter change from one state to another?"

The **three states of matter** are the solid state, the liquid state, and the gaseous (GAS-ee-uhs) state.

"Temperature. When water gets hot, it turns into a gas, and when it gets cold, it turns into a solid," Mike explained.

"That's a good explanation to start with," Dad said. "Instead of talking about temperature, though, scientists say that matter changes from one state to another because of a change in energy."

"A change in energy?" Mike repeated, puzzled. "What does energy have to do with temperature?"

"It has everything to do with it, because heat is a form of energy," Dad replied. "Now that you know about atoms, you can understand how a change in energy causes matter to change from one state to another.

"The first thing we need to understand is that atoms and molecules don't just sit around. They wiggle and jiggle and bounce off of each other and hit the things around them and jiggle some more. The energy that makes atoms and molecules move and jiggle is called heat energy, or **thermal energy**.

"When we say something has become warmer," Dad continued, "what we mean is that it has gained thermal energy. In other words, the atoms and molecules in the object have become more active. Atoms and molecules in hot objects have more thermal energy and move around more than atoms and molecules in cold objects.

**Thermal energy**, or heat energy, is the energy that makes atoms and molecules move and "jiggle."

FIGURE 3.1

To **expand** means to become larger.

To **contract** means to become smaller.

 Experiments #6-7

"Now, do you remember what Mom did the other day when Nick and Christie were full of energy? She sent them outside to play. Because they were so active, they needed more room.

"Energetic atoms and molecules are the same way," Dad explained. "The more they move, the more room they take up. The result is that matter **expands**, or becomes larger, when it is heated, and **contracts**, or becomes smaller, when it cools."[1]

Mike looked puzzled.

"Haven't you ever noticed before how heat makes things expand?" Dad asked. "It's time to head to our science lab, then. I don't think anyone else is planning to use the kitchen tonight."

Mike jumped up. "I'll beat you there!"

## 3.2 The Three States of Matter

"The experiments we did last night showed that matter expands when it's heated and contracts when it's cooled," Mike said. "But what does expanding and contracting have to do with the three states of matter?"

"When the atoms and molecules in a material are low in thermal energy," Dad explained, "they move and jiggle only a little. Since they aren't moving very much, they don't need much 'wiggle room,' and they are tightly packed together. In this case, the matter is a solid.

"If we add thermal energy to a solid, its particles become more and more active and start to take up more and more room. At a certain point, the particles become so energized that they move far away from their original places in the solid. They begin to slide

---

[1] The exception to this rule is when water freezes into ice. You will learn more about this in section 3.6.

over each other as they jiggle, and they move to new areas, instead of keeping the same neighbors that they had in the solid.

"This is called **melting**, and it is the process by which matter changes from the solid state to the liquid state. When a material is in the liquid state, its atoms and molecules slide over and around each other to fill up containers or spread over surfaces. For instance, when Nick spills his milk, it spreads out over the table, instead of sitting in a solid lump as it would do if it were a solid."

Mike burst out laughing at the thought of a solid lump of milk. "It sure would be easier to clean up. But imagine pouring solid milk into a cup, or trying to sip it up into a straw!"

"That would be very strange," Dad agreed, laughing. Continuing with the lesson, he asked, "Can you tell me what the third state of matter is?"

"The gaseous state," Mike said.

"That's right," Dad approved. "And how can we turn a liquid into a gas?"

"By boiling it," Mike said. "That's how we turn the water in a teapot into steam."

"Excellent! **Boiling** is one of the ways that a liquid can be turned into a gas," Dad said. "Matter boils when it has too much thermal energy to remain in the liquid state. Particles that used to slide around together as a liquid now have so much energy that they break away and float around by themselves.

**Melting** is the process in which a solid is changed to a liquid.

**Boiling** is a type of vaporization in which large amounts of a liquid change into a gas at once.

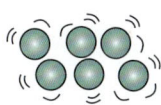

solid

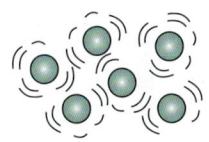

liquid

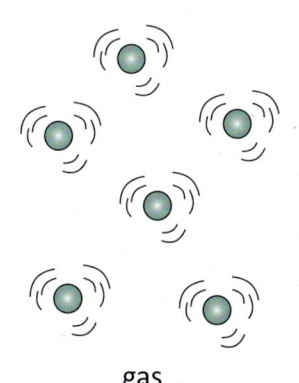

gas

FIGURE 3.2

27

**Water vapor** (VEY-per) is another name for steam.

**Vaporization** is the process of turning a liquid into a gas.

**Evaporation** is the type of vaporization that occurs when "speedy," energetic particles burst through the surface of the liquid and escape into the air.

**Condensation** is the process in which a gas is turned into a liquid.

**Freezing** is the process in which a liquid is turned into a solid.

"This is exactly what happens when water boils. The liquid water near the bottom of the pot becomes so hot that it rapidly expands into pockets of steam. These then bubble up to the surface of the water and are released into the air. Note that real steam, also called **water vapor**, is invisible. The 'steam' we see rising from a kettle of boiling water is really a cloud of tiny water droplets, and not a gas at all. These clouds are formed when real steam—that is, water vapor—is condensed into tiny drops of liquid water by the cooler air around the kettle.

"The process of turning a liquid into a gas is called **vaporization**," Dad explained. "Boiling is a type of vaporization, but it is not the only type. Liquids can also be turned into gases through **evaporation**. Boiling and evaporation are the two forms of vaporization.

"When water is boiled, large amounts of the water turn into water vapor at once," Dad explained, "but in evaporation, water turns into a gas little by little. For instance, if you leave a glass of water out on the counter, the water will slowly evaporate into the air. Even though you can't see the liquid water turning into water vapor, each day there will be less and less water in the cup, until eventually the cup will be completely empty."

"I've seen evaporation happen a lot," Mike said, "and I've learned that when it rains, a lot of the water in puddles evaporates back into the air. But how does evaporation actually work?"

"In a liquid, you remember, particles slide over and around each other as they jiggle," Dad explained. "The particles also bump into each other a lot, and when this happens, some particles are slowed down and some particles speed up. Evaporation occurs when 'speedy,' energetic particles burst through the surface of the liquid and escape into the air. Once this has happened millions of times, we begin to notice the water level going down.

"A big difference between boiling and evaporation is that boiling occurs only at certain temperatures, but evaporation can occur at almost any temperature. You have to heat water to 212°F before it will boil, but water evaporates even at room temperature. Of course, liquids evaporate faster at higher temperatures, because the extra thermal energy makes it more likely that some of the particles will gain enough energy to escape into the air."

"Is that why puddles of water disappear faster when it's sunny out?" Mike asked.

"Great example! I can see you really understand the three states of matter. Can you summarize the changes that a solid goes through on its way to becoming a gas?"

"Sure," Mike agreed. "It all goes back to how thermal energy makes matter expand. When you add thermal energy to a solid, the solid expands into a liquid. This is called melting. Then when you add thermal energy to the liquid, it expands into a gas, either by boiling or by evaporation. These are both forms of vaporization."

"Perfect," Dad praised. "Of course, you can also reverse the process. If you cool a gas far enough, it will lose thermal energy and become a liquid. For instance, water droplets form on the outside of a glass of ice water because the cold glass is cooling the water vapor in the air. This process is called **condensation**. If we cool the liquid even further, it will lose so much thermal energy that it will become a solid again. This process is called **freezing**. For instance, we can freeze liquid water into ice, or we can freeze sugar and cream into ice cream."

"Mmmm, I like the second option best!" Mike said.

## 3.3 Freezing and Boiling Points

You probably know that the freezing point of water is 32°F. When liquid water is chilled below this temperature, it turns into ice, and when ice is warmed above this temperature, it melts. The boiling point of water is 212°F. When water is heated above this temperature, it vaporizes as steam.

Of course, not everything freezes at 32°F and boils at 212°F. Gold remains solid until it is heated to 1948°F! Until it reaches that temperature, gold is "frozen solid," too cold for its atoms to slide over each other as a liquid.

Here are a few other freezing and boiling points:

Molten gold being poured into an ingot mold.

|  | Freezing Point | Boiling Point |
| --- | --- | --- |
| Water | 32°F | 212°F |
| Rubbing alcohol | -128°F | 181°F |
| Iron | 2800°F | 5181°F |
| Mercury | -38°F | 674°F |
| Oxygen | -362°F | -297°F |
| Lead | 621°F | 3180°F |
| Tungsten | 6177°F | 10,031°F |

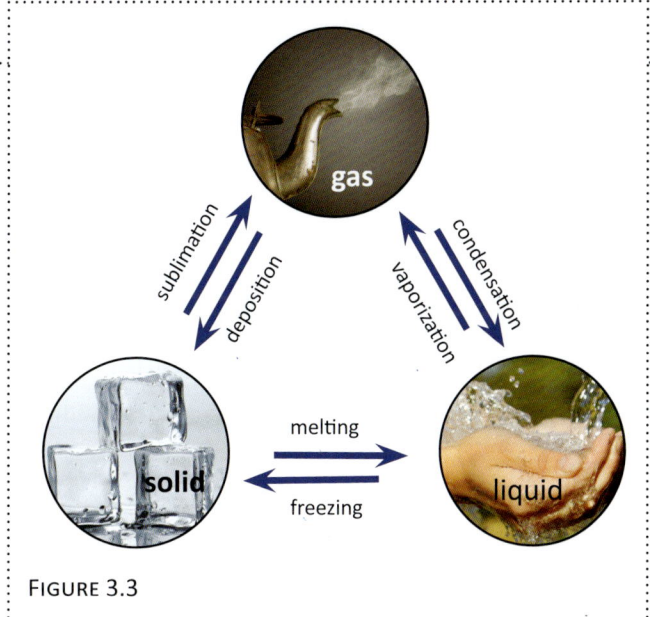

FIGURE 3.3

## 3.4 Sublimation and Deposition

In most cases, a solid must pass through the liquid state before it can become a gas. In certain situations, however, a solid can be converted directly into a gas. The process by which this happens is called **sublimation**. Dry ice, which is frozen carbon dioxide, is an example of a solid that undergoes sublimation. At room temperatures, dry ice sublimes into a gas without going through a liquid state.

The opposite of sublimation is **deposition**, and it is the process by which a gas turns directly into a solid. The frost that covers the ground on cold, winter mornings is the result of deposition. Frost is formed when water vapor—a gas—is converted directly into solid ice crystals, which decorate roofs, windows, plants, and ground with beautiful patterns.

**Sublimation** (suhb-luh-MEY-shuhn) is the process in which a solid is converted directly into a gas.

**Deposition** is the process in which a gas turns directly into a solid.

Charles D. Winters/Photo Researchers

 Workbook pgs. 9-10

 Experiments #8-9

31

Hot air balloons rise through the air because the air inside the balloon is warmer than the surrounding air. The air in the balloon is usually heated by a flame from a propane burner.

## 3.5  Warm Air Rises and Cold Air Sinks

You have learned that matter expands when it is heated, and contracts when it is cooled. Whenever a material expands or contracts, the density of the material is changed. This is because the same amount of matter is taking up more or less space than it did originally.

When matter expands, it becomes less dense, and when matter contracts, it becomes more dense. A cubic inch of liquid water is more than 1000 times denser than a cubic inch of water vapor, because the water molecules are packed together more tightly in the liquid state.

You have probably learned that warm air rises and cold air sinks, but do you know why? Since matter expands when it is heated and contracts when it is cooled, cold air is denser than warm air. There are more molecules of air in a square foot of cold air than there is in a square foot of warm air, so gravity "pulls" harder on cold air. Because of this, warm air floats on top of cold air in much the same way that oil floats on top of water.

## 3.6 Density of Ice

Ice is an exception to the rule that liquids contract when they are cooled. Most liquids freeze into solids simply by condensing and hardening. But when liquid water is frozen into ice, the water molecules arrange themselves in special patterns called ice **crystals**. Water molecules take up more room when they are arranged in ice crystals than they do when they are in a liquid state. Because of this, water expands when it freezes. As a result, ice is less dense than liquid water, so it floats.

If water didn't expand when it froze, ice would sink to the bottom of lakes and ponds in the winter. Little by little, the entire lake would fill up with ice, and the fish would freeze and die. By creating water so it expands when it freezes, God ensured that fish would have plenty of water in which to swim even in the winter.

**Crystals** are groups of molecules arranged in special patterns.

Experiment #10

## 3.7 Chemistry in a Nutty Carrot Cake

You have learned that heat makes solids expand into liquids and liquids expand into gases. So when Mike's mom puts a Nutty Carrot Cake in the oven, why does the heat make the liquid batter expand into a solid?

The expansion of cake batter in the oven is much more complicated than other types of expansion, because Nutty Carrot Cake is a mixture of many kinds of molecules. Some of its molecules, such as water, are very simple. Others, such as sugar and starch, are more complex.

**Proteins** are some of the largest, most complex molecules of all.

**Proteins** are some of the largest, most complex molecules of all. For example, hemoglobin, a protein in red blood cells, is made of 3032 atoms of carbon, 4816 atoms of hydrogen, 872 atoms of oxygen, 780 atoms of nitrogen, eight atoms of sulfur, and four atoms of iron, all linked together in exactly the right way. Nutty Carrot Cake doesn't contain hemoglobin, but the proteins in flour, milk, and eggs are just as complicated.

In fact, protein molecules are so complex that they don't react to heat the way simpler molecules react. If you heat an ice cube above its melting point, it will melt. But before protein molecules can be heated to their melting points, the heat causes them to change form and start acting in different ways.

For example, egg white is made of water and several different kinds of pure protein. Before an egg is cooked, the protein molecules are curled up in tiny balls, which stick together in a loose, gooey

34

mixture. When the egg is heated, the energy in the heat causes the protein molecules to "uncurl." In this new form, the individual protein molecules bond to each other very tightly, and the liquid egg white becomes solid. When this reaction occurs to the eggs in cake batter, it helps turn the batter into a solid.

Another important protein in cake is gluten. When flour is mixed with a liquid like water or milk, the molecules of gluten in the flour stick together in long, elastic strands. These strands form a net of stringy proteins, which hold the other ingredients together in a gooey batter or dough. When the batter is baked, the water evaporates and the gluten hardens. The result is a solid cake.

But why does the cake batter expand at all? Besides proteins, there are also less complex molecules in cake batter, and these do expand when they are heated. For instance, milk contains many water molecules, and baking powder produces carbon dioxide gases. When the water and carbon dioxide are heated, they expand to form bubbles of steam and gas. These bubbles are trapped by the elastic net of gluten, so when they expand from the heat, the rest of the cake expands, too. As the batter is baked and the water evaporates, the gluten solidifies around the bubbles. Then all Mom needs to do is take the large, fluffy cake out of the oven and let it cool!

Experiment #11

**Solubility** (sol-yuh-BIL-i-tee) is the ability of one material to be dissolved into another.

**Solutions** are mixtures that are made when one material dissolves into another.

Water is so good at dissolving things that it is called the **universal solvent**.

A **solvent** is a gas, liquid, or solid in which another material is dissolved.

When carbon dioxide is dissolved in water, the result is carbonated water, the basis for most soft drinks. The oxygen dissolved in water is what fish breathe through their gills.

## 3.8 Solubility

**Solubility** is another example of how the behavior of atoms and molecules determines how matter acts.

You have learned that the two main types of matter are pure substances and mixtures. You have also learned that there are two different kinds of pure substances, called elements and compounds. Have you noticed that there are different kinds of mixtures, too?

For example, Nutty Carrot Cake and sugar-water are both mixtures. When Mike looks at a slice of Nutty Carrot Cake, he can identify crunchy nuts, flecks of cinnamon, and pieces of shredded carrot. But when he examines a glass of sugar-water, he can't distinguish between the parts that are sugar and the parts that are water. The sugar has completely dissolved into the water.

Mixtures that are made when one material dissolves into another are called **solutions**. Many solutions are liquids, such as the mixture of sugar and water. Some solutions are gaseous, such as the mixture of oxygen and nitrogen in the air we breathe. Finally, some solutions are solids: a ruby is formed when a little bit of red chromium is dissolved into aluminum oxide, and a clear, hard lollipop is a solid solution of sugar, flavorings, and food colorings. Of course, solid solutions have to be mixed when the materials are in a liquid state, whether in a volcano (as in the case of the ruby) or on the stove (as in the case of the lollipop).

Although solutions can be made out of many different materials, water is so good at dissolving things that it is called the **universal solvent**. (**Solvent** is the scientific name for "dissolver," that is, the gas, liquid, or solid in which another material is dissolved.) Salt, sugar, alcohol, oxygen,

and carbon dioxide are just a few of the materials that dissolve in water.

When a material is able to be dissolved, we say that it is **soluble**. Sugar is soluble in water. Before sugar is poured into water, millions of sugar molecules are tightly "stuck" together in sugar crystals. When these crystals enter the water, they are pulled apart by molecules of water, which want to mix with the sugar molecules. The "pull" of the water molecules is stronger than the "stickiness" of the sugar molecules, so the sugar crystal falls apart and its molecules mix evenly with the water molecules.

When a material is able to be dissolved, we say that it is **soluble** (SOL-yuh-buhl).

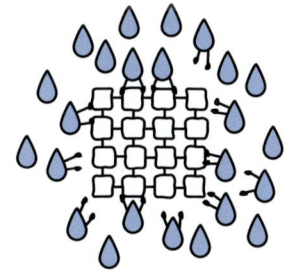 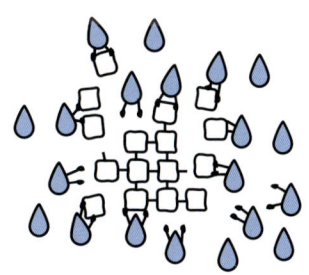

FIGURE 3.4  A sugar crystal as it first enters the water (left). A sugar crystal being "pulled" apart by water molecules (right).

Some materials are **insoluble** in water. This means they will not dissolve in water. For instance, if you drop a glass marble into water, it will not dissolve, because glass is insoluble in water. (It's a good thing, too—we wouldn't want our drinking glasses to dissolve when we fill them with water!) Unlike sugar molecules, the molecules in a glass marble "stick" to each other so tightly that the water molecules are unable to "pull" them apart.

Some materials are **insoluble** in water. This means they will not dissolve in water.

Many things which are insoluble in water will dissolve in other materials. For example, glass can be dissolved if it is placed in hydrofluoric acid, a powerful and dangerous chemical.

Hydrofluoric acid can dissolve glass because it has a stronger "pulling" ability than water does.

Do you remember the last time you sipped a glass of cold lemonade? I'm sure you drained the delightful mixture down to the last drop, and even used a spoon to scrape up the sugary mixture at the bottom of the glass. But wait! Since sugar dissolves in water, why did you find a layer of sugar crystals at the bottom of your glass?

Sugar is soluble in water, but only a certain amount. When no more sugar will dissolve, we say the water is **saturated**. Saturation occurs when the water molecules are so busy "holding on" to the

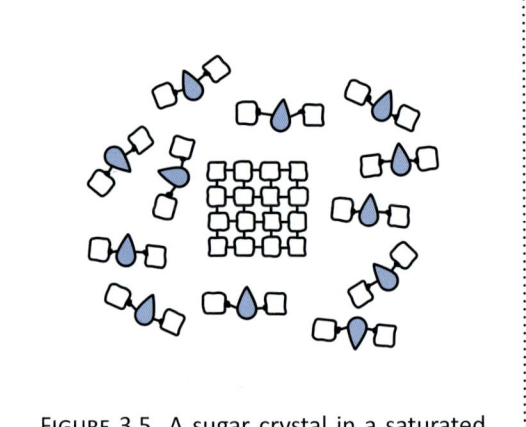

FIGURE 3.5 A sugar crystal in a saturated solution of water and sugar.

sugar molecules that have already dissolved that they no longer have the strength to pull more sugar crystals apart. Instead, the sugar crystals collect at the bottom of the glass.

The saturation point is different for different materials. A great deal of sugar can be dissolved into water, but only a little oxygen. The amount of a material that can be dissolved changes with temperature, too. Have you noticed that you can dissolve more sugar into warm water than into cold water? When Mike's older

When no more of a material will dissolve, we say that the solvent is **saturated**.

sister Daniella makes candy, she makes a smooth, thick syrup by dissolving several cups of sugar into just a little boiling water. Dani is able to do this because warm water molecules have more heat energy than cold water molecules. They use this extra energy to "pull" more sugar crystals apart.

## 3.9 The Scientific Method: Working in Less-than-Ideal Conditions

"The other day you discovered that the volume of Mom's Nutty Carrot Cake increased in the oven, even though she didn't add any mass to it. When you smashed it with your fork, its volume decreased again," Dad said.

"Right," Mike said. "That's because of the air bubbles."

"True. Of course, the air in the bubbles had mass, so when you smashed the cake with your fork to decrease its volume, you were also decreasing its mass a tiny bit."

"Oh. You mean my science experiment was wrong?"

"Not really. The air we breathe is made of oxygen and several other gases. As you have learned, the atoms and molecules in a gas are quite spread out. Remember that we defined mass as the amount of 'stuff' in an object? Since its atoms are so spread out, a square foot of air has very little mass compared with a square foot of Nutty Carrot Cake. So in your 'cake-smashing' experiment, it was all right to ignore the mass of the air. Ideally, of course, you should have taken the Nutty Carrot Cake into outer space, and made sure the bubbles were completely empty before smashing the cake with your fork," Dad teased.[2]

"But since we couldn't afford a trip to outer space, you had to do what most scientists do. You conducted your experiment and drew your conclusions as best you could, even though the conditions weren't perfect."

"You're not just saying that to be nice, are you? Real scientists actually work this way?" Mike asked.

"Yes, indeed. When I took biology classes in college, we had to write lab reports on the experiments we did. At the end of every lab report, we listed all the parts of the experiment that weren't perfect. For instance, I would write, 'The water flea I looked at under the microscope might have been sick,' or 'I might have made mistakes when I was recording the data,' or 'My ruler might have been slightly inaccurate.' Even when I thought the experiment had worked perfectly, I had to figure out reasons why my conclusions could be mistaken. Otherwise, my report would be incomplete. This is because scientists know that they are never working in perfect conditions, and they want to avoid making mistakes as they explore God's creation.

---

[2] Did you know there is no air in outer space?

"So if you were writing a lab report on your 'cake-smashing' experiment, you would be careful to mention the fact that you ignored the mass of the air bubbles. Then other scientists could take that fact into account when they read about your experiment. But you would conclude that the mass of the air bubbles probably didn't make much difference to the results of the experiment."

"You know, I wouldn't mind repeating the experiment more accurately some time," Mike observed. "Do you think Mom would help me out by baking another Nutty Carrot Cake?"

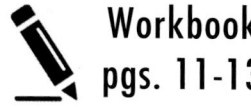

**Workbook pgs. 11-13**

# 4
## Chapter

# Types of Energy

The significance and joy in my science comes in those occasional moments of discovering something new and saying to myself, "So that's how God did it." My goal is to understand a little corner of God's plan.
— Henry "Fritz" Schaefer

**Energy** is the ability to do work.

**Work** occurs whenever a force moves an object to a new location.

To a scientist, playing hopscotch counts as work.

## 4.1 Energy and Work

"Dad," Mike exclaimed, running up to his father with science book in tow, "listen to this! My science book says, *'**Energy** is the ability to do work.'*"

"That's right," Dad affirmed.

"I don't see how that can be a good definition of energy," Mike said. "Nick and Christie have more energy than anyone I know, but they sure don't go around mopping the floors and washing the dishes all the time!"

Dad laughed. "Good point. But your science book is using the word 'work' in a different way than we use it in our everyday lives. A scientist says that **work** has been done whenever force is used to move an object to a new location. So from a scientific point of view, when Mom sends your brother and sister outside to burn off some energy by playing, she is sending them outside to work."

Seeing Mike's puzzled look, Dad continued. "Work occurs whenever a force moves an object to a new location. When Nick and Christie play, they are using force to move their arms and legs. So a scientist would say that when they chase each other, swing on the rope swing, and play hopscotch, they are doing work."

"I wish I 'worked' for a scientist," Mike grinned. "I think I'd like the chores he'd assign."

"Just because hopscotch counts as work to a scientist doesn't mean mowing the lawn doesn't," Dad warned. "In fact, you'd find that the scientific understanding of work would only help you out part of the time.

"To a scientist, work occurs whenever a force moves an object to a new location. This means that if there isn't any movement, there isn't any work. Remember last month when we helped Mom fill the new bookcase? We trooped back and forth from the garage, carrying the stacks of school books that she'd been storing in boxes. That counted as work because we were moving the books from the garage to the house. But when we stood still, holding a stack of books for her to load onto the shelves, a scientist would say that we weren't doing any work, because we weren't moving the books."

"That's not fair! It sure felt like work," Mike exclaimed. "And it took a lot of energy, which is the ability to do work."

"I know," Dad said. "But this energy wasn't used to move the books. Instead, the movement took place within your own body. While you held the heavy books, your cells absorbed food energy from your blood, your lungs took in oxygen, and your muscle cells expanded and contracted. By the end of the morning, you had used up so much energy that you needed to refuel your body with an apple and a granola bar, which your body then converted into more energy for your cells."

"So I *was* doing work?"

"In a way you were, but not the sort that can be measured

easily. Imagine trying to calculate the movement of each cell and molecule in your body! When scientists talk about work, they mean the sort of work that is done by tractors, baseball bats, hammers, pulleys, and moving arms and legs. But when they talk about the energy used by the cells and molecules in your body, they use terms like 'heat,' 'thermal energy,' and 'chemical energy.'"

"Oh no," Mike groaned. "Now I'm totally confused. What is the difference between 'thermal energy' and 'chemical energy'?"

"I think you'll be learning about these terms in your science book soon," Dad said. "For now, just remember that work occurs whenever a force moves an object to a new location."

## 4.2  Mechanical Energy

**Mechanical energy** is the type of energy that is used to do work. For instance, mechanical energy is present in the turning of the wheels on a bicycle or the moving of a pair of scissors. It is also present when Nick throws a ball and when Christie skips rope. This type of energy has to do with moving objects.

**Mechanical energy** is the energy that is used to do work.

Mechanical energy gets its name because it is the type of energy we see in moving machines. But mechanical energy is at work in much more than just machines. When a river flows over a waterfall, it is using mechanical energy. When Mike mows the lawn and when his sister Dani stirs milk into flour to make biscuits, they are using mechanical energy. You are using mechanical energy right now to flip the pages of your science book.

## 4.3 Thermal Energy

**Thermal energy** has to do with heat. In fact, the word "thermal" comes from the Greek word for "heat." When we say that an object is hot, we mean that it has a lot of thermal energy. The more thermal energy an object has, the faster its atoms and molecules are moving and jiggling. The atoms and molecules in a steaming mug of cocoa are moving and jiggling very quickly, so it has a lot of thermal energy. In other words, it is hot.

**Thermal energy**, or heat energy, is the energy that makes atoms and molecules move and jiggle.

When the atoms and molecules in an object are moving slowly, we say the object is cold, or has only a little thermal energy. An ice cube does not have much thermal energy, because its atoms and molecules aren't moving very much.

When we say something has become warmer, we mean that the thermal energy of the material has increased. In other words, the atoms and molecules in the material have become more active.

## 4.4 Chemical Energy

**Chemical energy** is released when atoms and molecules react with each other to form new materials. For example, the molecules in baking soda and vinegar react with each other to produce carbon dioxide gas. This reaction doesn't just produce carbon dioxide—it also releases chemical energy. In fact, the baking soda-vinegar reaction releases so much energy that it can be used to launch toy rockets, inflate balloons, and explode model volcanoes.

**Chemical energy** is the energy that is released when atoms and molecules react with each other.

Another example of chemical energy is the process of cellular respiration that goes on in the cells of our body. Respiration is the chemical process that we use to produce the energy we need to move and grow. In the process of respiration, glucose, a type of sugar, reacts with oxygen in order to release water, carbon dioxide, and—you guessed it—energy. We use this energy to work, play, and study.

Chemical energy is also released when fireworks are exploded or when a gun is fired. Gunpowder is made of sulfur, charcoal, and potassium nitrate (saltpeter); when these materials are heated by friction or fire, they react with each other to release huge amounts of chemical energy in the form of high-power explosions.

## 4.5 Energy Transformation

"I'm not sure I understand the difference between all the types of energy, Dad," Mike observed.

"I'm not surprised you find it challenging," Dad said. "Different types of energy can look very similar, especially when they turn into one another."

"What do you mean, 'turn into one another'?"

"Chemical energy can turn into thermal energy, and mechanical energy can turn into chemical energy, and so on. You've seen this happen all the time.

"For example," Dad continued, "consider the reaction between baking soda and vinegar. If you place the baking soda and vinegar in a bottle and fit a cork in the top, the reaction between the baking soda and vinegar will shoot the cork out of the bottle just as

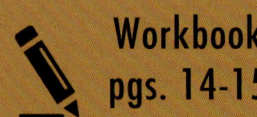

Workbook pgs. 14-15

if it were a rocket. In other words, the chemical energy of the reaction will be converted into the mechanical energy needed to lift the cork into the air.

"Thermal energy can also be converted into mechanical energy. You have learned that materials expand when they are heated. In this process of expansion, thermal energy is converted into mechanical energy. For instance, a one-mile-long length of steel will expand by about three feet if it is heated from 70°F to 100°F. Since bridges are made out of steel, engineers build them with expansion joints or other mechanisms, so they can expand in hot weather without cracking or bending.

"Mechanical energy can be converted back into thermal energy, too. For instance, it is possible to start a fire by rubbing one stick against another. The friction, or rubbing, between the two sticks transforms mechanical energy into thermal energy. This is one method people used to start a fire before the invention of the match. You can experience the production of heat from motion for yourself by rubbing two wooden pencils across each other for a few seconds.

"The energy needed to light a match undergoes several transformations. When you strike a match against a matchbox, the mechanical energy of your moving hand is converted into thermal energy by friction. This thermal energy causes a reaction between the chemicals on the match head and the chemicals on the matchbox. In turn, the chemical energy from this reaction produces so much thermal energy that the wooden matchstick bursts into flame. These flames produce even more thermal energy, which we use to light a candle, start a wood fire, or explode fireworks."

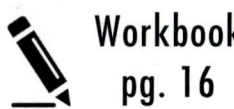
Workbook pg. 16

## 4.6 Potential and Kinetic Energy

When we speak about mechanical, thermal, and chemical energy, we are classifying energy by how it is produced and used. Mechanical energy is seen in the motion of large-scale objects, thermal energy comes from the movement of tiny particles, and chemical energy is released by chemical reactions.

We can also classify energy by whether or not it is currently being used. For example, when Nick and Christie play tag, the energy they have is being put to immediate use. But when Nick and Christie are sitting still working math problems, they have the energy to play tag, but they are not currently using it.

Energy that could be used, but isn't being used at the moment, is called **potential energy**. When Nick holds a ball above his head, the ball has potential energy. All he has to do is let go of the ball and the ball will fall to the ground.

Energy that is being used is called **kinetic energy**. The ball's movement towards the ground is an example of kinetic energy. As the ball falls towards the ground, its potential energy is transformed into kinetic energy.

Another example of potential energy is the energy that is stored in baking soda and vinegar. Until we combine the baking soda with vinegar, the energy present in the baking soda is potential energy, because it is not being used. As soon as we combine the

**Potential energy** is energy that is not being used, but could be.

**Kinetic** (ki-NET-ik) **energy** is energy that is being used.

two chemicals, the potential energy is converted into kinetic energy, or energy that is being used. This kinetic energy makes the mixture bubble, fizz, and expand.

## 4.7 More about Potential Energy

An object has potential energy because of its position. When a basketball player holds the ball over his head, he is holding it in a position that gives it potential energy, because the force of gravity will pull the ball to the ground as soon as he lets go. The farther the ball is from the ground, the more potential energy it has, because it will move farther and faster when it is released.

The potential energy in baking soda also comes from its position. Baking soda is a compound, which means that it is made of millions of baking soda molecules. Each of these molecules is made of six atoms—a carbon atom, a hydrogen atom, a sodium atom, and three oxygen atoms—bound together in a particular arrangement. Baking soda has potential energy because its molecules are arranged in these positions. All we have to do is add vinegar, and the atoms in the baking soda molecules will change their positions, turning their potential energy into kinetic energy.

 Workbook pg. 17

 Experiment #12

## 4.8 Thermal Energy and the Sensation of Heat

Mike had just finished mowing the lawn, and was cooling his face against a glass of ice-cold lemonade. "Mom, why does this glass feel cold?" he asked.

"I thought you learned about ice when you studied how heat makes things expand," his mom said. "And didn't you learn about thermal energy this week?"

"Yes," Mike affirmed. "I know that when things are hot, their atoms and molecules jiggle a lot, and when things are cold, their atoms and molecules jiggle only a little. This jiggling is called thermal energy. But why do objects *feel* hot or cold when I touch them?"

"Oh, I see what you are asking," his mom said. "You want to know what is happening when our senses detect that something is hot or cold.

"Well, the first thing we need to realize is that thermal energy—heat—always moves from an object that is hotter to an object that is colder. So when we touch a cold glass, the thermal energy in our fingers immediately starts leaving our fingers and moving to the glass, lemonade, and ice. When our fingers tell us something is cold, what they are really indicating is that thermal energy is leaving our hands.

"On the other hand, if we curl our fingers around a mug of hot cocoa, thermal energy flows out of the warm mug and into our hands. Our hands feel warmer because they are gaining thermal energy."

"So when I touch a cold glass and my hands start feeling colder, it's not because 'coldness' is flowing into my hands?" Mike asked.

FIGURE 4.1 Thermal energy always flows from the warmer object to the cooler object.

"Not at all," Mom said. "What we call 'warmth' or 'coldness' is simply the feeling we have when thermal energy enters or leaves our fingers. When you feel that an object is hot or cold, it's not because you are sensing some invisible substance called 'warmth' or 'coldness.' It's because your fingers detect that thermal energy is entering or leaving your body."

"So my fingers are little 'energy meters'?" Mike laughed.

Mom smiled. "Exactly!"

## 4.9 Conduction

"Wait a minute—" Mike paused. "How does thermal energy move from my hand into the glass?"

"Another good question! Thermal energy is the jiggling of atoms and molecules, so one of the main ways thermal energy moves around is by atoms and molecules bumping into each other. This is called **conduction**.

"When you touch a cold glass, the warm, jiggling molecules in your skin come in contact with the molecules of the glass, which are moving only slowly. The energetic molecules in your skin 'bump into' the molecules in the glass, making them jiggle faster. These glass molecules then bump into other glass molecules, which bump into other glass molecules and the molecules in the lemonade, until all the molecules in the glass of lemonade are jiggling faster than they were before.

"Meanwhile, the molecules in your skin move more slowly after they bump into the slow glass molecules. They have transferred some of their 'jiggle' energy to the molecules in your glass, so they no longer have as much energy themselves. Eventually, the amount of energy in your hand and in the glass becomes equal, and the glass no longer feels hot or cold to you."

FIGURE 4.2

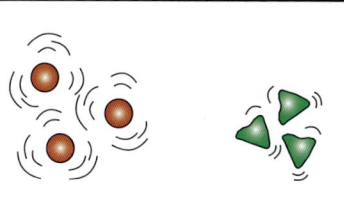

No transfer of thermal energy will occur through conduction unless the objects are brought into direct contact.

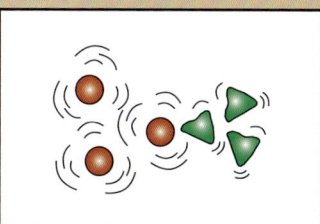

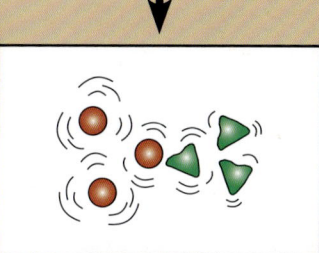

Each particle passes on some of its "jiggle" to its neighbors.

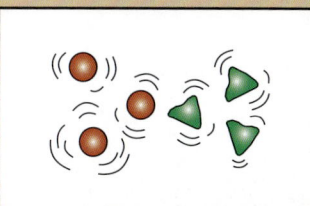

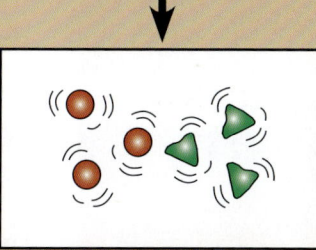

Conduction ends when the particles in both objects have the same amount of "jiggle," or thermal energy.

**Conduction** is the transfer of thermal energy through atoms and molecules bumping into each other.

**Insulators** are materials that don't "pass on" thermal energy very well.

**Conductors** are materials that are particularly good at conducting thermal energy.

**Workbook pg. 18**

**Experiment #13**

## 4.10 Insulators and Conductors

Mike thought for a few seconds. "Conduction doesn't happen all the time, though, does it? If I wear gloves when I touch the cold glass, it doesn't feel cold. This means that thermal energy isn't leaving my hand. Why doesn't the glove pass the thermal energy from my hand to the cold glass?"

"Good point!" Mom said. "The molecules in the glove do not absorb much thermal energy when the warm molecules in your hand bump into them. It is simply difficult for the glove's molecules to start jiggling when other molecules bump into them. And of course, since the glove doesn't absorb thermal energy itself, it can't pass on thermal energy to the glass.

"Materials that don't 'pass on' thermal energy very well are called **insulators**. Plastics, fabrics, wood, and air are good insulators. This is why potholders are made out of fabric. The molecules in the fabric of the potholder do not easily absorb and transfer thermal energy. Potholders and other insulators 'block' the transfer of thermal energy.

"On the other hand, some materials are particularly good at conducting thermal energy. These materials are called **conductors**. Metals are the best conductors, because their molecules easily start jiggling when they are 'bumped into' by other molecules. This is why a metal spoon gets hot so quickly when you dip one end into a pot of boiling water. The end of the spoon that is immersed in the water immediately starts jiggling from the thermal energy in the water, and this jiggling quickly spreads throughout the whole spoon as one metal molecule bumps into another."

"Hey, let's go try it!" Mike exclaimed.

## 4.11 Convection

"Mom, yesterday you taught me about conduction and how molecules and atoms pass on thermal energy by bumping into each other," Mike said. "But this can only happen when two objects are right next to each other. Can objects transfer thermal energy if they aren't right next to each other? It seems like they must be able to, or else I'd have to touch a heater to feel warmer."

"You're right," Mom said. "For the particles in one object to bump into the particles in another, there has to be direct physical contact. So if you stop touching a cold glass, thermal energy will no longer be transferred from your hand to the glass, at least not by conduction.

"There is a 'long-distance' method for objects to transfer thermal energy, though. This method is called **convection**. Convection is the way thermal energy is spread throughout liquids and gases, just as conduction is the way thermal energy is transferred between solids," Mom explained.

"Does convection occur in a glass of lemonade?" Mike asked.

"Yes it does," Mom said. "We only talked about conduction yesterday, but your glass of lemonade was also an example of convection. Before I explain convection, can you remind me how conduction works in a glass of lemonade?"

"Sure," Mike agreed. "The warm molecules in my hand were jiggling faster than the cold molecules in the glass. So when I touched the glass, the warm molecules 'bumped into' the molecules in the glass and made them jiggle faster. Then the molecules in the glass bumped into other glass molecules, making *them* jiggle faster, until the thermal energy from my hand was spread throughout the whole glass," Mike explained.

**Convection** is the transfer of thermal energy by means of circulating liquids and gases.

"Or at least, that's what would have happened if I had held the glass long enough," Mike added. "I didn't want the glass to 'suck' all the heat out of my hand, so I stopped holding it after a while."

"Once the glass became warmer, what happened to the lemonade inside? Did it remain cold?" Mom asked.

"No, it became a lot warmer," Mike said. "That's why the ice started melting."

"So the glass transferred thermal energy from your hand to the lemonade, which means it was acting as a conductor, not an insulator," Mom said. "This isn't hard to understand now that we've learned about conduction. But what we want to know is what happened once the thermal energy in the glass was passed on to the lemonade.

"Unlike the molecules in a solid," Mom explained, "molecules of lemonade are free to move around in a glass. The particles in a liquid, remember, are spaced much farther apart than the particles in a solid. So when the molecules of lemonade that are touching the glass become warm, they don't transfer heat by bumping into other molecules. Instead, they move away from the side of the glass, and colder molecules take their place. When these molecules are warmed up, they move aside too, and new molecules take *their* place. This process is called convection, and it occurs until all the lemonade is warmer."

"Wow!" Mike exclaimed. "But how do the molecules know that they should move over when they have gotten warmer?"

"They use miniature cell phones to communicate with each other," Mom explained, keeping a perfectly straight face. Mike stared at her in disbelief.

"Just kidding!" she laughed. "Convection is an amazing example of how God created the different properties of matter to work together. Convection occurs because almost all materials are more dense when they are cold than when they are warm. Do you remember why this is so?"

"Because materials expand when they are heated?" Mike guessed.

"Exactly. One cup of hot water actually has less matter in it than one cup of cold water, because water expands when it is heated. Not as many water molecules fit in a cup when they are hot as when they are cold," Mom said.

"Now, do you remember why an object that is denser than water sinks, but an object that is less dense floats?" Mom asked.

"Not really," Mike admitted sheepishly. "I found that chapter in my science book kind of hard to understand."

"Then this is a great chance to review. When an object is denser than water, it sinks. When an object is less dense, it floats. This is because the force of gravity is greatest for the object that is most dense, or has the most matter in it," Mom explained.

"For instance, if we pour a cup of hot water into a cup of cold water, being careful not to mix them into warm water, gravity will be stronger for the cold water than for the hot water, because the cold water is denser. As a result, the cold water will sink to the bottom, and the hot water will rise to the top."

"All right, I think I understand now," Mike said. "How does this work in convection?"

"When the particles of lemonade next to the glass are warmed, they gain thermal energy and move farther apart. They are now less dense than the cold lemonade around them, so they float up to the top of the glass, and cold molecules take their place. These molecules are also warmed, become less dense, and float to the top of the glass."

Mike was impressed. "That is awesome! I wouldn't have thought of that in a million years!"

"I wouldn't have either," Mom agreed. "Isn't our Creator amazing?"

Convection occurs in gases as well as in liquids. When a heater is placed in a cold room, the air that is touching the surface of the heater gains thermal energy through conduction. The warm air rises towards the ceiling, and cold air takes its place. The moving air forms a convection current that warms the entire room.

## 4.12  Radiation

Experiment #14

You have learned that thermal energy can be transferred by conduction or convection. Thermal energy can also be transferred through **radiation**, which is a way of transferring heat and light through electromagnetic waves. Electromagnetic waves are beyond the scope of this book, but they are not beyond the scope of your experience. Rays of sunshine, X-rays, radio waves, and microwaves are all types of electromagnetic waves. Electromagnetic radiation from the Sun is what provides the Earth with light and warmth. When you bask in the warm sunshine on a chilly day, you are being warmed by the thermal energy carried by electromagnetic waves.

**Radiation** is a way of transferring heat and light through electromagnetic waves.

The Sun sends out—or radiates—many different types of electromagnetic waves, including X-rays, rays of visible light, and infrared rays. **Infrared rays** are the invisible waves of energy that are radiated by an object such as a hot clothes iron. Unlike the Sun, a clothes iron does not become hot enough to glow, so it does not radiate visible light. A hot iron radiates infrared rays, though, and if you place your hand near a hot iron you can feel these as warmth.

Electromagnetic waves, including infrared rays, transfer thermal energy by "bumping into" the molecules in objects. For example, when infrared rays from the hot iron strike your hand, they "bump into" the molecules in your fingers and skin. The molecules in your hand absorb the energy in the infrared ray, so they begin to jiggle more vigorously, and your hand becomes warmer. The infrared rays have transferred thermal energy from the iron to your hand.

It makes sense for the Sun or a hot iron to radiate infrared rays. But did you know that *you* are also radiating infrared rays? Every object in the world is radiating infrared rays, even icebergs. This is because every object, no matter how cold it feels, contains at least a small amount of thermal energy. Even the molecules in an ice cube are jiggling a little bit, so the ice cube must be radiating some energy in the form of infrared rays. The reason we don't notice the energy radiated by an ice cube is that an ice cube is so cold in comparison with other objects.

**Infrared** (in-fruh-RED) **rays** are the invisible waves of energy that are radiated by an object such as a hot clothes iron.

A heater warms a room by radiation as well as by convection.

Workbook pgs. 19-21

# Chapter 5

# Electricity and Magnetism

The beauty of electricity . . . is not that the power is mysterious . . . but that it is under law, and that the taught intellect can even govern it largely.

— Michael Faraday

## 5.1 Static Electricity

"Ouch!" Mike exclaimed. "That was quite a shock! All I did was touch the doorknob."

"You've been scuffing your feet, haven't you?" Dad observed. "Rubbing your shoes across the carpet increases the effects of **static electricity**."

"Static electricity . . ." Mike repeated. "Oh, I remember! You told us about static electricity when we were sticking the balloons to the ceiling on Christie's birthday!"

"I'm glad you remember that! The static electricity that you felt when you touched the doorknob came from rubbing your shoes on the carpet. In the same way, the static electricity that made the balloons stick to the ceiling came from rubbing the balloons on your clothing or hair."

"Or someone else's hair," Mike grinned. "Dani really appreciated the 'electric hair-do' we gave her!"

"It was rather 'shocking,' wasn't it?" Dad laughed. "Do you know why her hair frizzed up like that?"

"Not really, but the same thing happened to Nick the other day when he went down the tunnel slide at the park. When he came out, his hair was standing on end!"

"I would like to have seen that," Dad laughed.

**Static electricity** refers to the build-up of electrons (negative charge) on objects.

"In both cases you were seeing the effects of static electricity. Let's find out how static electricity makes these things happen."

## 5.2  Positive and Negative Charge

"Do you remember learning that atoms are made out of protons, electrons, and neutrons?" Dad asked.

"Not very well," Mike admitted. "Except that they're really tiny!"

"That's for sure!" Dad agreed. "Protons and neutrons are tiny particles that form the center, or nucleus, of an atom. Electrons are even smaller particles that are constantly whizzing around the nucleus. Atoms are so small that a sheet of plastic wrap is about 100,000 atoms thick. But protons are 100,000 times smaller than an atom, and no one even knows how small an electron is!"

"My head hurts!" Mike exclaimed.

Dad smiled. "Their small size certainly makes electrons difficult for scientists to study! But it doesn't make them unimportant. On the contrary, electrons are at the heart of static electricity."

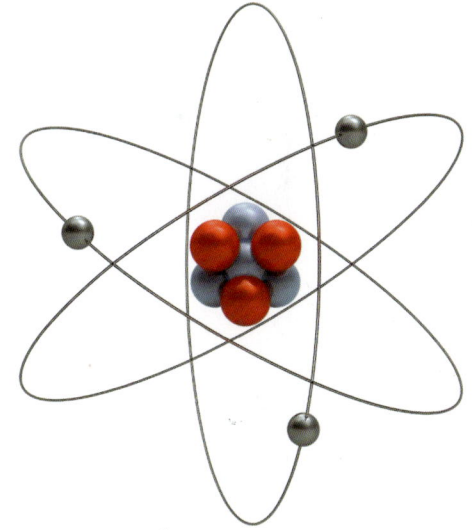

"So electrons cause sparking doorknobs, sticky balloons, and frizzy hair?" Mike asked.

"Exactly. Here's how they do it. Normally, the atoms making up a material contain the same number of electrons as protons. So if I tell you that a balloon contains three billion protons, you know that it also contains three billion electrons.[1] At least, it usually does.

"When you rub a balloon on your head, some of the electrons in your hair are transferred to the balloon. The balloon now

---
[1] A balloon actually contains many more electrons and protons.

The minus sign (−) is the symbol for a negative charge.

The plus sign (+) is the symbol for a positive charge.

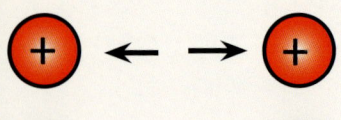

Like charges repel.

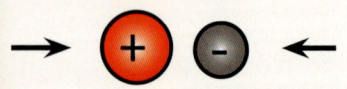

Unlike charges attract.

FIGURE 5.1

An object that is **negatively charged** has more electrons than protons.

An object that is **positively charged** has more protons than electrons.

has more electrons than protons, so we say it is **negatively charged**. Similarly, your hair now has more protons than electrons, so we say it is **positively charged**.

"The most important thing to know about positive and negative charges is that **unlike charges attract and like charges repel**. 'To repel' means to push away, so if two positive charges or two negative charges are brought near each other, they will push each other away. But what happens if a positive charge is brought near a negative charge?" Dad asked.

"They will pull towards each other?" Mike guessed.

"Absolutely."

"Is this what makes balloons stick to the ceiling?" Mike asked.

"Your brain's really working! The attraction, or 'stickiness,' between unlike charges is precisely what makes negatively-charged balloons stick to the ceiling. After you rub a balloon in your hair, the balloon is negatively charged because some of the electrons in your hair have jumped over onto the balloon. And you know what happens when you bring something with a negative charge near something with a positive charge."

"They move towards each other!" Mike exclaimed. "So when the balloon sticks to the ceiling, it's because the balloon has a negative charge and the ceiling has a positive charge."

"That's almost right. When you rub a balloon in your hair, your *hair* becomes positively charged because it gives some of its electrons to the balloon," Dad explained. "This is why Dani's hair stuck to the balloon when you were giving her an 'electric hair-do': her positively-charged hair was attracted to the negatively-charged balloon.

"The balloon sticks to the *ceiling* for a slightly different reason. Unlike Dani's hair, the ceiling as a whole is not positively charged, because it still contains the same number of electrons and protons. So we wouldn't expect the balloon to be attracted to it.

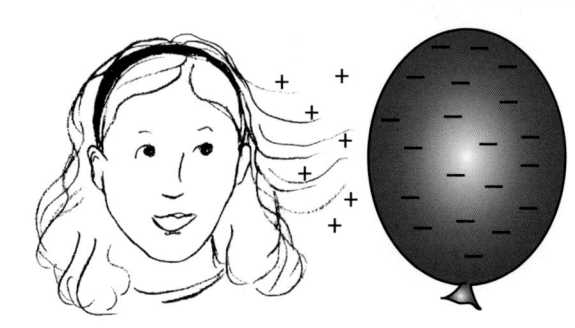

FIGURE 5.2 Dani's positively-charged hair is attracted to the negatively-charged balloon.

"But when the balloon comes near the ceiling, the electrons in the balloon repel the electrons in the ceiling. These electrons don't leave the ceiling, but they move away from the surface. Even though there is an equal number of electrons and protons in the ceiling as a whole, the surface of the ceiling now has more protons. This means the surface of the ceiling is positively charged, so the negatively-charged balloon sticks to it!"

"Now that's what I call cool!" Mike exclaimed. "I can't wait for a chance to explain static electricity to Nick and Christie. Maybe I can persuade Dani to help me demonstrate the attraction between her hair and a balloon."

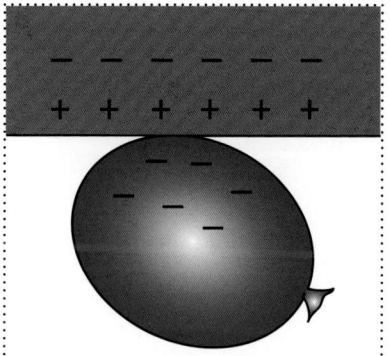

FIGURE 5.3 The negatively-charged balloon sticks to the positively-charged surface of the ceiling.

"I suspect she will 'repel' your suggestions," Dad punned. "You might want to have Nick demonstrate instead by going down the tunnel slide again."

"Yes, that should work . . ." Mike paused. "In that case, the electrons in Nick's hair will be rubbed off onto the plastic tunnel. Then the tunnel will be negatively charged and his hair will be positively charged. So of course, the tunnel and his hair will be attracted to each other, and Nick's hair will stand up straight!"

Mike paused as he thought of an objection. "But the other day when we were at the park, Nick's hair kept standing up straight even after he left the tunnel. Why didn't it immediately lie flat again? The electrons in the tunnel couldn't have been holding it up any more."

"To answer that question, let's remember the most important fact about charges: **unlike charges attract and like charges repel**. Was the charge in each of Nick's hairs like or unlike the charge in the rest of his hair?"

"It was like," Mike said. "All his hairs had a positive charge."

"So was each hair attracted to the other hairs or was it repelled by them?"

"Repelled," Mike affirmed.

"That's why Nick's hair stood on end. Each of his hairs was trying to get as far away as it could from every other hair."

"His hairs were scared of each other!" Mike laughed. "Now when someone's hair stands on end, I can ask, 'Did your hair have a scare?'"

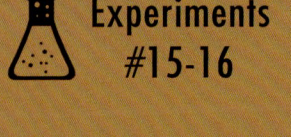
Experiments #15-16

## 5.3 Static Electrical Discharge

"Dad," Mike observed, "I've learned how static electricity can make a balloon stick to the ceiling. But I don't understand how static electricity caused the spark when I touched the doorknob."

"Actually, it's similar to what happened with the balloon and the ceiling. When you rubbed the balloon in your hair, lots of electrons were transferred from your hair to the balloon. This left the balloon with more electrons than protons, so it had a negative charge.

"The same thing happens when you scuff your shoes across the carpet. The rubber and plastic in the soles of your shoes collect lots of electrons from the carpet. Since electrons can't travel through the air, they are trapped on your body. This gives you a negative charge, just like the negative charge on the balloon.

A **static electrical discharge** occurs when "extra" electrons in a negatively-charged object "jump" through the air into another object.

"Of course, since you are larger and heavier than a balloon, you don't feel a 'pull' between your body and the wall! The electrons feel a 'pull,' though. As soon as you get close to something that is positively charged, or is connected to something that is positively charged, the electrons in your hand 'jump' through the air into the other object. This is called a **static electrical discharge**."

Mike jumped up and started scuffing his shoes across the carpet. "Here goes!" he exclaimed, approaching the door. Dramatically, he touched his finger to the doorknob. Nothing happened.

"Hey, why didn't it spark?" Mike asked, disappointed. "Do I need to scuff my shoes harder?"

"Actually, the amount you scuff your shoes across the carpet doesn't make as much difference as the weather," Dad explained. "Yesterday when the doorknob shocked you it was dry and cold, but today it's raining. Remember that I said the electrons get trapped in your body because they can't travel through the air? Well, on damp days there is a lot of water vapor in the air, and electrons *can* travel through water vapor. So on damp or humid days like today, the electrons escape into the moist air instead of being trapped in your body."

"The next time we have dry weather, can we have a 'scuffing and sparking' party?" Mike asked.

"Actually," Dad said, "you don't need to scuff your shoes to have a 'spark party.' Electrons move from the carpet into your shoes even when you stand still. Scuffing your feet just speeds up the process by allowing your shoes to touch more of the carpet. So if the day is dry enough, you can collect enough electrons to have a 'spark party' just by dragging your feet gently."

"Does that go for balloons, too?" Mike queried. "Will electrons be transferred if I just touch a balloon to Dani's hair?"

"Yes. But of course, it happens faster if you rub."

"And speed is important when it comes to giving Dani an 'electric hair-do'!" Mike grinned.

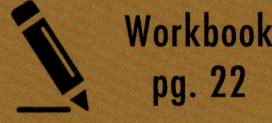

Workbook pg. 22

## 5.4 Magnetism

Have you ever played with magnets? There are probably some on your refrigerator right now, holding grocery lists, artwork, and photos to the door. A **magnet** is a piece of metal that can attract, or pull, other metals. When an iron nail is brought near a magnet—zip!—the magnet pulls the nail towards itself.

The force that magnets use to pull other metals is called **magnetism**. This force can seem magical because it makes things happen from a distance. In reality, magnetism is much like gravity, which also pulls on things from a distance.

Gravity exerts a force on all mass, but magnetism only affects materials like iron and steel. These materials are called **magnetic materials**. Other materials, such as copper, glass, paper, and wood, are not affected by magnetism at all, so they are called **non-magnetic materials**.

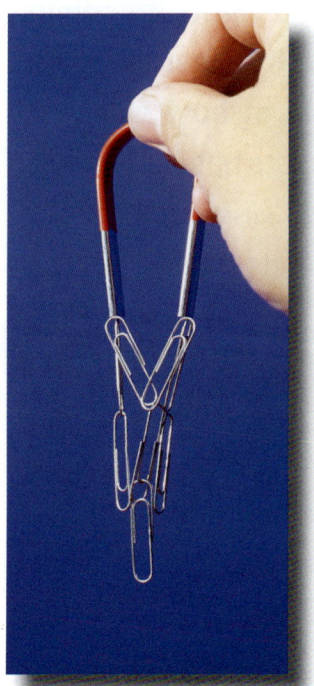

When magnetic materials are touched to a magnet, they start acting like magnets themselves. Other magnetic materials are attracted to them just as if they were weaker versions of the permanent magnet. This is called **induced magnetism**, and it is the reason you can hang a long string of paper clips from a magnet even though only one or two are actually touching the magnet. The paperclips have become temporary magnets. They will lose their magnetism as soon as the permanent magnet is removed.

Andrew Lambert /Photo Researchers

A **magnet** is a piece of metal that can attract, or pull, other metals.

**Magnetism** is the force that magnets use to pull other metals.

**Magnetic materials** are materials that are attracted to magnets.

**Non-magnetic materials** are materials that are not attracted to magnets.

**Induced magnetism** is the type of magnetism that magnetic materials have when they are in contact with a magnet.

Unlike poles attract.

Like poles repel.

Unlike charges attract.

Like charges repel.

If you have ever played with magnets, you know that they not only attract magnetic materials—they attract other magnets, too. But magnets can also repel, or push away, other magnets. It all depends on which ends of the magnets are being brought together. When you hold two strong magnets in one position, it might take all your strength to keep them from flying towards each other. But if you turn one of the magnets around, all your strength might not be enough to force them together!

Magnets act this way because every magnet has two ends, or poles: a north pole and a south pole. If you suspend a bar magnet from a string, the north pole is the one that points north after the magnet has come to rest, and the south pole is the one that points south. If you bring the north pole of one magnet near the south pole of another magnet, the two magnets will be attracted to each other very strongly. They might even jump out of your hands towards each other! But if you bring a north pole near a north pole or a south pole near a south pole, the two poles will repel each other with great strength. This is because **unlike poles attract and like poles repel**.

If you think this rule sounds familiar, you're right! It's almost identical to the rule for electricity that says **unlike charges attract and like charges repel**. This is one example of how electricity and magnetism are connected. In fact, scientists have discovered that electricity and magnetism are actually a single force which sometimes acts electrically and sometimes acts magnetically.

## 5.5 Magnetic Fields

As you have learned, a magnet is a piece of metal that attracts or repels other magnets and magnetic materials. Scientists have discovered that magnets attract and repel materials according to strict rules.

Have you noticed that the strength and direction of a magnet's "pull" changes depending on where you hold the piece of magnetic material? For example, when you hold a paperclip half an inch away from a magnet, the magnetic attraction is quite strong. But if you move the paperclip several inches away, the magnetic attraction becomes much weaker.

Since magnets attract and repel other materials according to strict rules, scientists are able to calculate the exact strength and direction of a magnet's "pull" at any position. They do this by mapping out an invisible force field around the magnet, called its **magnetic field**. This is the region in which the magnet's "pull" can be felt by magnetic materials. The magnetic field of a magnet on your refrigerator is very small and weak—often it can hold only a few pieces of paper to the refrigerator. The magnetic field of a large, horseshoe magnet is much larger and stronger. A nail or paperclip can be nearly a foot away from such a magnet and still be attracted to it. And some magnets are so strong that they are used in junk yards to move iron scraps and even entire cars!

A **magnetic field** is the region in which a magnet's "pull" can be felt by magnetic materials.

A junk-yard magnet is a special type of magnet called an electromagnet, because it can be turned on and off with an electric current. If a junk-yard magnet were an ordinary magnet, there would be no way to release the iron scraps once they were attached to the magnet. Electromagnets are another example of how electricity and magnetism are connected.

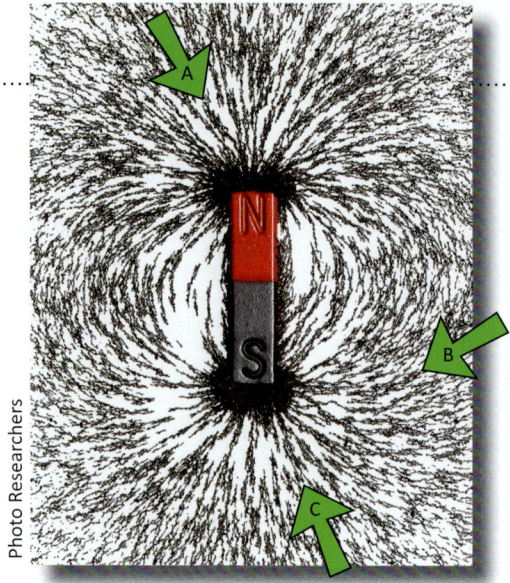

FIGURE 5.4

Even though magnetic fields are invisible, there are many ingenious ways to make them visible. For example, if we sprinkle iron filings over and around a magnet, the tiny bits of metal will jump into position and form a map of the magnetic field. Do you see the lines of iron filings in Figure 5.4? These lines indicate the direction of the magnet's force at any particular point.

For example, if you drop a paperclip at point A, it will experience a force in the direction of the "A" arrow. If you drop a paperclip at point B, it will experience a force in the direction of the "B" arrow. And if you drop a paperclip at point C, it will experience a force in the direction of the "C" arrow. Do you see how many iron filings have clumped around the north and south poles of the magnet? This is because the magnetic field is always strongest around a magnet's poles.

Magnetic fields can also be used to show the attraction or repulsion between two magnets. Can you figure out which of the photos below shows magnetic attraction and which shows magnetic repulsion?

Experiment #17

## 5.6 Electric Fields

"What is this a picture of?" Dad asked Mike, pointing to Figure 5.5.

"A magnetic field," Mike declared confidently, remembering the experiments he had done with iron filings and a magnet.

"Nope," Dad grinned. "It's an **electric field**."

"You're asking me a trick question!" Mike objected. "I've never even heard of electric fields before."

"Well now you have! But I don't blame you for saying it was a picture of a magnetic field; magnetic and electric fields look very similar. The big difference, of course, is that one is produced by magnets and one is produced by the electric charge in protons and electrons."

Mike studied the picture more closely, comparing it to the photos of magnetic fields in his science book. "There are some differences in the pictures, though. For a magnet, the field lines are in loops that come out of the north pole and return into the south pole. In the picture of the electric field, the field lines are coming out of the black spot in straight lines, not loops."

"Brilliant!" Dad praised, thoroughly impressed. "I never would have noticed that so quickly if I were first learning about electric fields. You're right—magnetic field lines are arranged in loops, but electric field lines flow out of the electron or proton in straight lines. That's because electrons and protons don't have north and south poles. For a magnet, there is an attraction between its north pole and its south pole, so the lines of force in its magnetic field flow between its poles. But there isn't any attraction between the two sides of an electron or proton, so the field lines just flow straight out. If you bring another proton near a proton, it will be

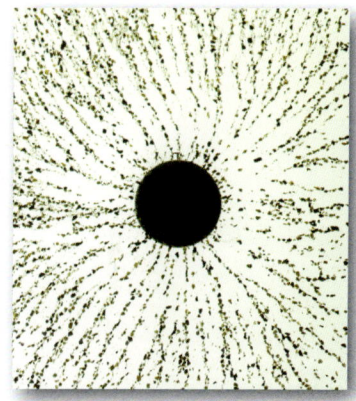

Ted Kinsman/Photo Researchers

FIGURE 5.5

An **electric field** is the region in which the "push" or "pull" of an electron or proton can be felt by other particles.

repelled straight out. And if you bring an electron near the proton, it will be pulled straight towards the proton.

"The concept of electric fields gives us another way of imagining static electricity. When a negatively-charged balloon is attracted to the ceiling, it's because the electrons in the balloon are attracted to the protons on the surface of the ceiling. The attraction between the balloon and the ceiling is an electric force field, which we can map out by drawing electric field lines between the balloon and the ceiling. These electric field lines allow us to calculate the direction of the electric field at any point between the balloon and the ceiling."

"The direction of the electric field?" Mike repeated.

"Yes. For example, the electric field lines in Figure 5.6 tell us that if we dropped a proton at point A, it would be repelled by the ceiling and attracted towards the balloon in the direction of the 'A' arrow. If we dropped an electron at point B, it would be repelled by the balloon and attracted towards the ceiling in the direction of the 'B' arrow. This is how electric fields can help us understand static electricity."

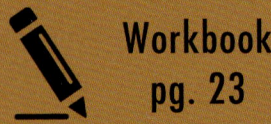

Workbook pg. 23

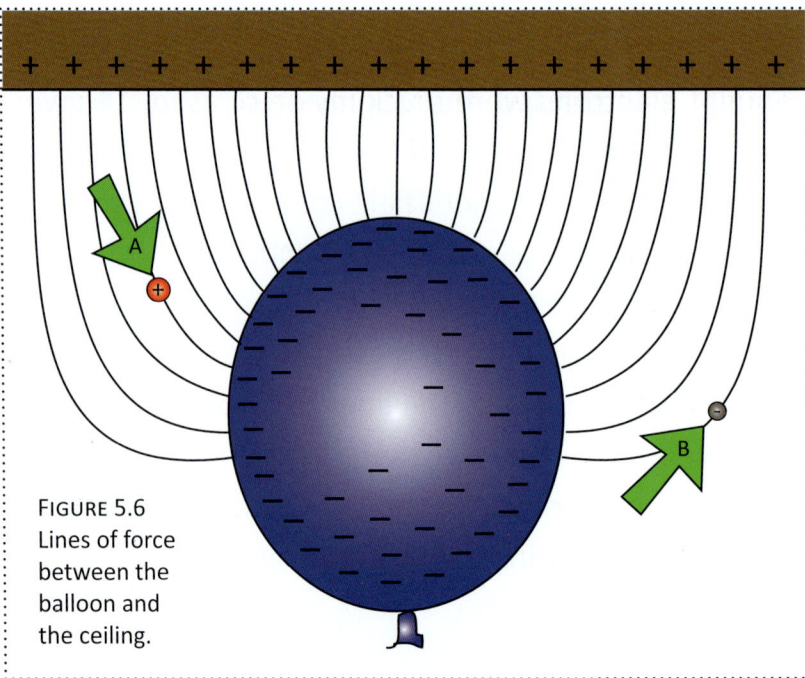

FIGURE 5.6
Lines of force between the balloon and the ceiling.

## 5.7  Electric Current

"What happens when I turn an electric light on and off?" Dad asked a few days later, flipping the light switch up and down.

"When you turn the switch on, electricity flows through the wires and lights up the bulb," Mike explained. "When you turn the switch off, the electricity stops, so the light stops, too."

"That's a good explanation to start with," Dad said. "But what exactly is 'flowing' through the wires? What is electricity?"

"I'm not really sure," Mike admitted.

"Don't worry," Dad reassured Mike. "For hundreds of years, even the most brilliant scientists had no idea what electricity really was. In fact, for a long time they thought it was a kind of invisible liquid! That's why we still talk about electricity as 'flowing' through wires, as if it were water flowing through a hose.

"When we plug a toaster, lamp, or fan into an electrical outlet, a stream of millions of electrons begins flowing through the electrical wire. This flow is called an **electric current**."

"I thought electrons were stuck inside of atoms. How can they move around inside the wire?" Mike asked.

"That's a good question! Many materials 'hold onto' their electrons very tightly, and it is very difficult to get an electric current to flow through them. But other materials, especially metals, 'hold onto' their electrons only loosely, and the electrons can 'hop' from one atom to the next. This is why we use a metal wire to connect the light bulb to the electrical outlet instead of using a rubber hose!"

An **electric current** is the flow of millions of electrons through an electrical wire.

Mike laughed. "I bet I could make a song about that! *Water flows through rubber hoses, but electrons flow through metal wires*," he chanted.

"Hey, we're not done learning about electricity, yet," Dad interrupted, with feigned severity. "You're supposed to ask what makes electrons move from one end of a wire to the other."

"What makes electrons move from one end of a wire to the other?" Mike asked willingly.

"Electrons are set in motion when a generator or a battery acts as a pump to 'push' them along the wire. You see, the energy that an electric current provides is not free—we have to put energy into the generator or battery in order to get energy out.

"In a certain sense, an electric current is like a current of water. Think of the fountain at the park. Small streams of water flow out and fall into the pool beneath it. When the water is at the top of the fountain, does it have potential or kinetic energy?"

Mike thought for a few seconds, and then said, "Potential energy."

"That's right. At the top of the fountain, the water has the possibility, or potential, to fall into the pool, but it isn't doing it yet. As soon as the water starts falling into the pool, its potential energy turns into kinetic energy, or energy that is being used.

"When the stream of water hits the pool, its kinetic energy is used to produce splashes and ripples in the pool. Then it comes to rest and floats around with the rest of the water. At this point, it has little or no energy, so it won't 'do' anything if left to itself.

"But of course, it isn't left to itself. A pump, hidden in the base of the fountain, sucks the water up and pushes it to the top of the fountain. By lifting the water to the top of the fountain, the pump gives some of its energy to the water. The water reaches the top with a new supply of potential energy, which turns into kinetic energy when the water falls, and is then used up producing splashes and ripples. The pump then pushes the water to the top of the fountain once more, which gives it the potential energy to begin the whole process over again."

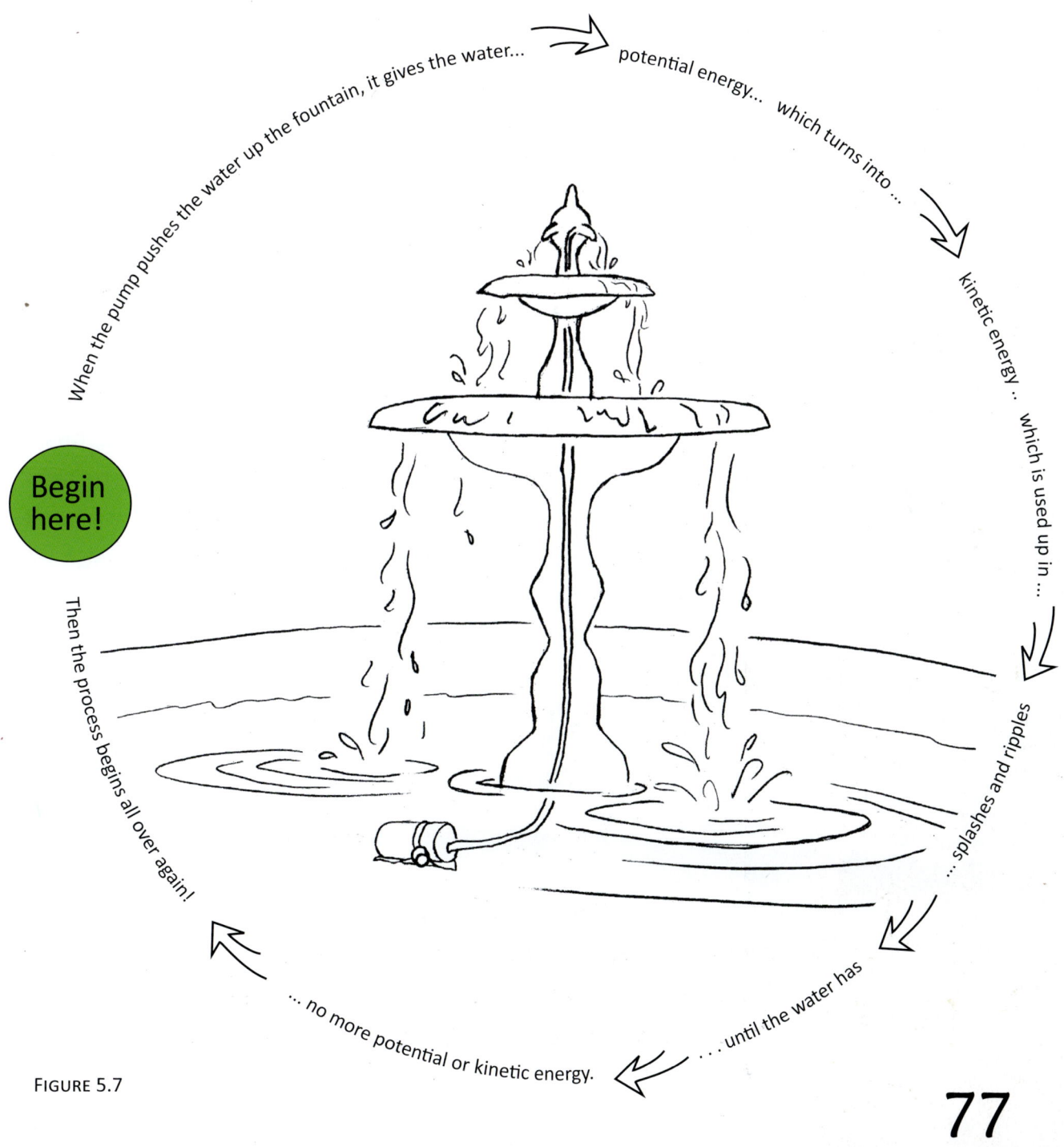

FIGURE 5.7

"And this is how an electric current works?" Mike asked, puzzled.

"I know it sounds crazy to compare the current in a fountain with the electric current in a lamp," Dad admitted. "But they really are similar in the sense that neither current will do anything unless some sort of 'pump' gives it energy.

"Left to itself, the water in a pond will just sit there or drift around randomly. But if the water is given energy by a pump, it can splash and make ripples, turn a toy waterwheel, and rinse the dirt off Nick's fingers. In the same way, the electrons in a wire will just sit still or move aimlessly if they are left to themselves. But when a battery or a generator gives them energy, the electrons can turn a fan or light a bulb."

"So batteries and generators are 'electron pumps'?" Mike asked.

"In a way, yes. Generators and batteries 'push' on the first electrons in the wire. These electrons 'push' the next electrons, which 'push' the electrons next to them, and so on, until the 'push' reaches the light bulb. The light bulb converts this 'push' of energy into light and heat.

Doug Martin/Photo Researchers

"Once the electric current has passed through the light bulb, it no longer possesses much energy. Its energy has been used up by the light bulb, just as the energy of the stream of water was used up by making splashes and ripples. The electrons would just sit around like the water in the pond at this point, except that the battery or generator pump is 'sucking' them back along the wire. And when the electrons return to the battery or generator pump, they are given a new 'push' of energy which begins the process all over again," Dad concluded.

**Workbook pg. 24**

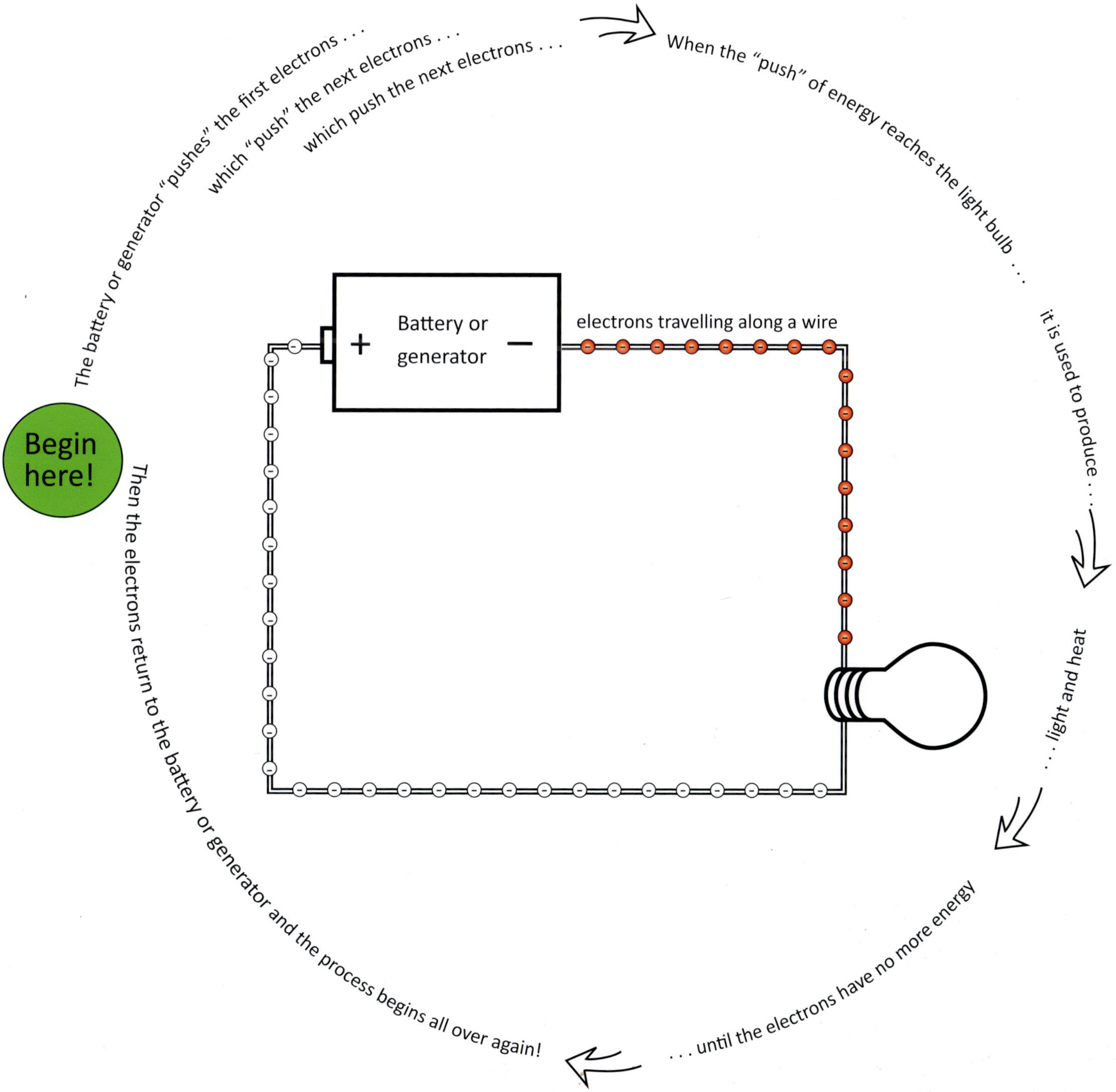

Begin here!

The battery or generator "pushes" the first electrons... which "push" the next electrons... which push the next electrons... When the "push" of energy reaches the light bulb... it is used to produce... ...light and heat ...until the electrons have no more energy Then the electrons return to the battery or generator and the process begins all over again!

Battery or generator

electrons travelling along a wire

FIGURE 5.8

## 5.8 Electrical Potential Difference

"How does a battery or generator 'push' on electrons?" Mike asked. "I thought they were too tiny even to see."

"Good point," Dad replied. "Batteries and generators can't push electrons through the wire physically, the way a pump pushes water through a pipe. Instead, batteries and generators make use of the attraction between electrons and protons. Do you remember our discussion of static electrical discharge? What made the electrons in your hand 'jump' through the air towards the door handle?"

"My hand was negatively charged because I had picked up lots of extra electrons by scuffing my shoes on the carpet," Mike explained. "Since opposite charges attract, the extra electrons in my hand 'jumped' through the air as soon as I came near something that wasn't negatively charged."

"Exactly," Dad said. "Scientists would describe that situation by saying that there was an **electrical potential difference** between your hand and the door knob. In other words, there was a difference in charge, because your hand was negatively charged while the door knob either was positively charged or was connected to something that was positively charged.

"Batteries and generators use electrical potential difference to 'push' electrons through the wire. Batteries and generators generally have one positive end filled with protons and one negative end filled with electrons, which means there is an electrical potential difference between the two sides. When the two ends of a battery or generator are connected by a wire, the electrons in the negative end 'push' against the electrons in the wire because they want to get closer to the protons in the positive end. The attraction between the two sides of the battery

---

An **electrical potential difference** is a difference in charge between two materials. In other words, one material is negatively charged and the other material is positively charged, or is connected to something that is positively charged.

An object that is **positively charged** has more protons than electrons.

An object that is **negatively charged** has more electrons than protons.

or generator is what keeps the electrons moving through the wire."

"The generator must push the electrons along the wire really quickly," Mike commented. "If I flip a switch on one side of the room, the light on the other side turns on instantly!"

"Actually, in an ordinary wire the electrons move only about four inches per hour!" Dad said. "But the 'push' of energy from the generator or battery moves through the wire almost instantly, and it is this 'push' that makes the light turn on.

A generator in an electrical power plant

"The process is similar to how the energy of a 'squeeze' travels down a tube of toothpaste. When you press on the toothpaste molecules on one end of the tube, these push on the next and so on until the paste on the other end is squeezed out.

"Notice that the *energy* of your squeeze moves almost instantly from one end of the tube to the other. You don't have to wait a while for your 'squeeze' to reach the other end of the tube and squirt out toothpaste.

"But the toothpaste itself hardly moves through the tube at all—maybe only a quarter of an inch. In the same way, the 'push' of electrical energy travels down the wire in less than a second, but the electrons themselves move very slowly."

Mike shook his head. "I'm going to have to read through this chapter in my science book a couple of times before I'll be able to understand it."

81

"Don't worry if you find it confusing at first," Dad said. "You will study electricity again in future science courses. The most important thing for you to remember right now is that all our electrical equipment—lamps, fans, refrigerators, ovens, computers, cell phones—is powered by a flow of energy passed from electron to electron. This energy isn't produced out of thin air: it has to be 'pumped' into the wires by batteries or generators, which in turn are powered by work that is done by the motion of water, wind, chemicals, or other forces."

"So if we get electrical energy out of one end of a wire, we know that there was work done on the other end to provide us with this energy?" Mike asked.

"Exactly," Dad affirmed. "In other words, **we can't get something for nothing.** At least not in science, that is."

Mike looked puzzled, and then exclaimed, "Oh, I know what you mean! We can't get something out of nothing in science, but God can get something from nothing. He created the world out of nothing."

"Very true. And even in our everyday lives some things seem to be created from nothing. For instance, your mom and I have loved you since before you were born. That's more than eleven years now! Where do we get all that love? Don't you think our store of love is going to run out?"

> The fact that "we can't get something for nothing" is called the **Law of Conservation of Energy.**

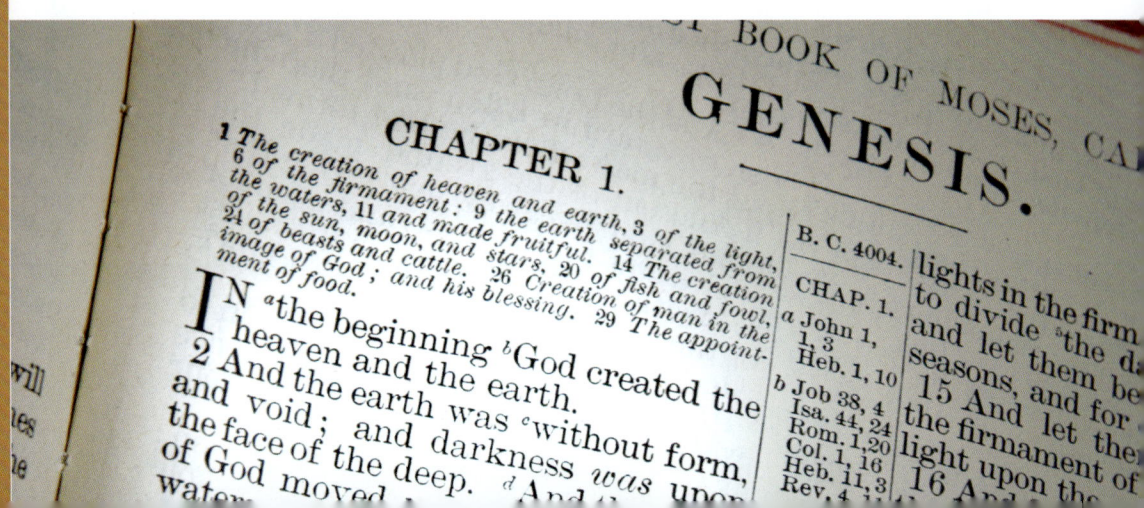

"No," Mike laughed, puzzled. "I know you and Mom will always love me."

"Indeed we will! In fact, our store of love keeps getting bigger and bigger, as if it were being mysteriously produced out of nothing. That's the way love works. The more people give love, the more love they have to give. Do you think this might be one of the ways we were made in the image of our Creator, who created everything from nothing out of love for us?"

## 5.9 Measuring Electricity

Sometimes you will hear special terms used when talking about electricity. "Coulomb," "ampere," "volt," and "watt" are all terms that describe how much electricity is present or at work. These terms were named after famous scientists who made breakthroughs in the study of electricity. Three out of these four major scientists were Catholics: Charles Coulomb, Andre-Marie Ampere, and Alessandro Volta.

Charles-Augustin de Coulomb

- **"Coulomb"** is a term that is used to measure how much charge—how many charged particles—an object has. One coulomb is equal to the charge of six billion, billion electrons (six quintillion or 6,000,000,000,000,000,000).

One **coulomb** (KOO-lom) is equal to the charge of six billion, billion electrons.

The **ampere** (AM-peer) is a unit that tells us how many electrons are going past a certain point every second.

The term **"volt"** measures the strength of the "push" that a generator or battery gives to an electric current.

The term **"watt"** measures how much electrical power a certain appliance uses.

- The **ampere** is the unit that we use to measure an electric current. It tells us how many electrons are going past a certain point every second. One ampere stands for one coulomb per second. When you turn on a 60-watt light bulb on a 120-volt line, the wire carries half an ampere. In other words, three billion, billion electrons—half a coulomb—are passing a certain point every second. If this sounds fast, it's only because we often imagine that the electrons in a wire are lined up in single file. On the contrary, electrons are so small that there

Andre-Marie Ampere

are millions and billions of them simply in the width of the wire. They don't have to move very quickly for three billion, billion of them to go by every second. This is why Mike's dad said that electrons move only about four inches per hour.

- **"Volt"** is the term we use to measure how great the electrical potential difference is between the two sides of a battery or generator. Thus, this term measures the strength of the "push" that a generator or battery gives to an electric current. An ordinary AA battery provides a "push," or voltage, of 1.5 volts. The electricity in your home is "pushed" by huge generators at electrical power plants. When the electricity enters your home it has a voltage, or push, of 120 volts. No wonder Mike's mom is so careful to keep Nick's and Christie's fingers away from the electrical outlets!

Alessandro Volta

- The term **"watt"** measures how much electrical power a certain appliance uses. A 60-watt light bulb on a 120-volt line uses half an ampere of current. This means that when you turn the light on, it is using a current of three billion, billion electrons per second, and this current has a voltage, or "push," of 120 volts. Some appliances use a lot of electrical power per second, and others use much less. Surprisingly, Dani's hair dryer uses 1875 watts of electrical power, while her dad's laptop computer uses only 50 watts.

James Watt

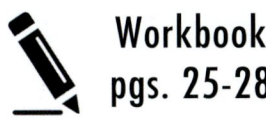
Workbook pgs. 25-28

Experiment #18 (optional)

# 6
## Chapter

# Simple Machines

The perplexity of life arises from there being too many interesting things in it for us to be interested properly in any of them.
— G.K. Chesterton

## 6.1 Simple Machines

"Mike, have you ever wondered why human beings rule the planet instead of animals?" Mom asked. "Lions and elephants are a lot stronger than we are."

"But we're smarter than they are," Mike said. "We can invent weapons and tools to make us stronger than animals."

"That's exactly right!" Mom said. "God created human beings in His own image, giving us the gift of reason and the ability to make rational choices. No other creature can solve problems and plan for the future as well as human beings.

"One of the ways man has used his intelligence is to invent machines to multiply the force of his muscles. Can you think of some examples?"

"In the old days farmers had to harvest crops by hand," Mike said, "but now they have tractors to plow and harvest their fields. Machines have also made travelling a lot easier. Instead of walking, we can drive a car or fly in a plane."

"Great examples! And don't forget about the machines we have in our own home, like the washing machine, the vacuum cleaner, heaters, and fans."

"And the dishwasher," Mike added. "That's one of my favorites! It sure took longer to wash the dishes by hand, that time the dishwasher broke."

"Yes," Mom smiled, "the way to appreciate anything is to do without it for a while."

"That's sort of like the G.K. Chesterton magnet we have on the refrigerator, isn't it?" Mike commented. "The one that says, *'The way to love anything is to realize that it might be lost.'*"

"You caught me!" Mom laughed. "Yes, I was paraphrasing Chesterton. In his writings he always emphasizes the surprising beauty and goodness of Creation."

"Why does he think the beauty of Creation is surprising?" Mike asked.

"Because none of us could have imagined it," Mom explained. "Chesterton calls Creation 'the best of all impossible worlds,' and says that if we saw the world for the first time, we would think it was fairyland. If life seems boring sometimes, it just means we have started to take the people and things around us for granted."

"We sure don't take things for granted in science!" Mike commented.

"No, we don't," Mom said. "Science really begins when someone is curious enough to ask questions about something everyone else has been taking for granted. And that's what we're going to do during the next two weeks. Instead of taking machines for granted, we're going to explore how they really work.

"Now, dishwashers, vacuum cleaners, and the other examples we've mentioned are complex machines. All machines, even the most complex, are combinations of six simple machines that men have used for centuries. The six simple machines are the inclined plane, the screw, the wedge, the lever, the wheel and axle, and the pulley. We're going to study them one at a time, so let's get started with the inclined plane."

## 6.2 Inclined Plane

If you have stairs in your house, go right now and run up and down them.

Are you out of breath? This is because you were doing work. Work is done whenever a force moves an object to a new location, and when you climbed the stairs, your muscles were using force to move your body up the stairs.

It took effort to climb the stairs, of course, but imagine how hard it would be to get up to the second floor of your house without stairs. You would have to jump straight up! Not even Mike's dad could do that. Thankfully, most two-story houses come with a simple machine called the **inclined plane**. In other words, they come with stairs.

An **inclined plane** is a simple machine that allows us to raise objects little by little instead of all at once.

An inclined plane is a simple machine that helps us move things from a low place to a high place with much less effort. Stairs are a good example of an inclined plane.

A ramp is also an inclined plane. Perhaps you have helped a neighbor carry boxes and furniture into a moving van, or maybe your own family has moved recently. I hope the moving van came with a ramp, because it is much easier to lift heavy furniture little by little along a ramp, instead of lifting it straight up in the air all at once.

Inclined planes make it easier to lift heavy objects because they allow us to move them little by little. Lifting a washing machine straight up into a moving van uses a very large force, but if you move it up a ramp, you only have to put out—or exert—a little force with each step.

On the other hand, when you use an inclined plane to lift an object, you have to move the object farther. An eight-foot ramp is longer than the three feet from the ground to a truck bed. So even though you are exerting less force *at a time* as you carry the object up the ramp, the total amount of force you use is the same as if you had lifted the object straight up. An inclined plane does not make a job require less force; rather, it allows us to exert the necessary force little by little instead of all at once.

## 6.3 Screw

The **screw** is a special type of inclined plane. To understand this, we first need to realize that inclined planes do not have to be straight. When Mike's family travels to visit his grandparents in the mountains, they drive on twisting roads that wind up the face of the mountain. These roads are not straight like a ramp, but they are still inclined planes. They allow Mike and his family to ascend the mountain a little at a time, instead of all at once like mountain climbers.

A spiral staircase works the same way. This type of staircase doesn't look like a ramp, but it still does the work of an inclined plane. It allows us to move up and down a little at a time instead of all at once.

Screws are also types of inclined planes. If you were smaller than an ant, a screw would look like a spiral staircase. You could climb up it and down it, and it would make it easier for you to raise yourself to the dizzying height of two inches.

Of course, you aren't smaller than an ant, so even though screws look like miniature spiral staircases, we use them for different purposes. One of the main reasons screws are useful is that they can convert a back-and-forth, rotational movement into an

Miracle Staircase at Loretto Chapel, Santa Fe, NM

The **screw** is a special type of inclined plane that can convert a rotational movement into a vertical movement.

up-and-down movement. For example, when you remove the cap from a bottle with a screw-top lid, you twist it sideways. But the bottle-cap doesn't move sideways—it moves upwards! The rotational movement of your hand has been converted into a vertical movement. This is one of the ways we use simple machines: to change the direction of a force.

Screws can also be used to hold things together. Carpenters use screws instead of nails when they want to fasten things together very tightly, because the threads—or ridges—of a screw grip the wood. It is virtually impossible to pull a screw straight out of a board, even with a crowbar. The only way a screw can be removed is by unscrewing it—that is, by twisting it horizontally. In other words, the only way to move a screw in a vertical direction is to provide a horizontal, rotational force.

## 6.4 Wedge

The **wedge** is a simple machine made out of two inclined planes placed back to back. Like a screw, a wedge can change the direction of a force. A screw converts horizontal, rotational movement into vertical movement, and a wedge converts vertical movement into horizontal movement.

For example, when Mike's dad uses a hatchet to split firewood, his goal is to divide the log into two pieces. Even though he moves the hatchet vertically, the movement he is really interested in is the horizontal movement of the log splitting in two. Since the head of his hatchet is shaped like a wedge, its sharp edge sinks into the

The **wedge** is a simple machine made out of two inclined planes placed back to back.

wood easily. As he forces the wedge farther into the material, the blade gets wider and wider, pushing the wood apart until it splits.

If Mike's dad were as strong as a giant, he could split firewood by pulling the log apart with his bare hands, just as you can pull apart string cheese. Wedges are useful because normal human beings are not strong enough to exert that much force at once. A wedge allows us to push the log apart little by little, just like the sharp edge of a knife allows us to push a knife into a carrot little by little. (Have you ever tried to cut a carrot with the blunt edge of a knife?)

Wedges aren't always used to split things. Sometimes we use them to make room for one object inside another. The sharp tip of a nail is a wedge that pushes wood out of the way to make room for the rest of the nail.

The prow of a boat works the same way. The pointed shape of the prow acts as a wedge to slice through the water, making room for the rest of the boat.

 **Workbook pg. 29**

## 6.5 Lever

Like the inclined plane, the **lever** is a simple machine that makes it easier to move heavy things. A lever is a long board or rod that rotates around a fulcrum. A **fulcrum** is a sturdy object or surface that the board rotates around. When you place a heavy object on one side of a lever, you can lift the object by pushing down on the other side of the lever.

In Figure 6.1, the fulcrum is set exactly in the middle of the lever. This lever does not change the amount of effort required to lift the load, but it does change the direction of the force. When you press down on one side of the lever, the other end of the lever moves up. Your downward force has been converted into an upward force.

The conversion of a force from downward to upward is especially useful when it allows you to use your body weight to lift the load. This is how a teeter-totter works. If you try to lift a friend straight over your head, you probably won't succeed. But if you are both sitting on a teeter-totter, you can lift your friend easily because you are using your body weight instead of just your muscle strength. The teeter-totter makes your job easier by changing the direction of your force.

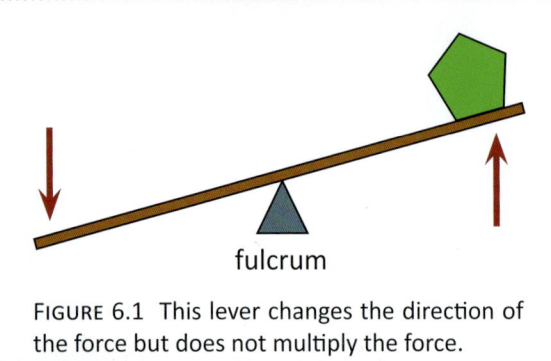

FIGURE 6.1 This lever changes the direction of the force but does not multiply the force.

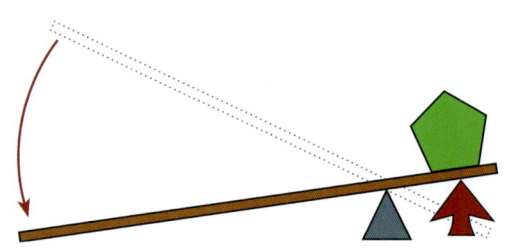

FIGURE 6.2 A small downward force is converted into a large upward force.

In Figure 6.2, the fulcrum is positioned close to the load instead of being in the middle of the lever. This type of lever changes the direction of our force, and it also multiplies our force. When we press down on the long side of the lever, we are able to lift the load with much less effort. The closer the fulcrum is to the load, the easier it is to lift the load.

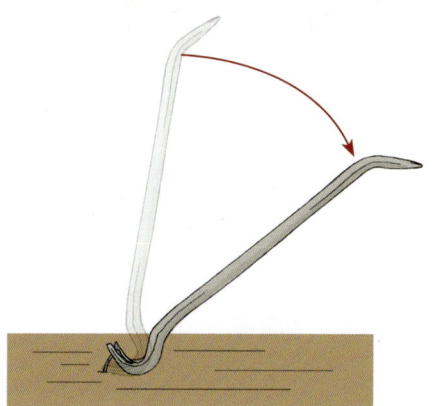

A crowbar is a good example of a lever that multiplies our force. The "crook" in a crowbar is the point around which the crowbar rotates, so it acts as a fulcrum. The fulcrum is positioned near the prying end of the crowbar, and the end of the crowbar that we pull on is quite long. This shape multiplies our force so we can pry out nails more easily.

A **lever** is a long board or rod that rotates around a fulcrum.

Even though we say that a lever "multiplies our force," it is important to realize that a lever does not actually produce more force. It takes less force *at a time* to remove a nail with a crowbar than it does to pull it straight out, but we have to apply the force over a longer distance (see Figure 6.3). So we must either use a very large force over a short distance, or a small force over a long distance. In the end, we do the same amount of work.

A **fulcrum** is a sturdy object or surface that the board rotates around.

FIGURE 6.3

**LARGE FORCE** x short distance   =   small force x **LONG DISTANCE**

When we want to convert a small force into a large force, we position the fulcrum so it is nearer to the load than to ourselves, as in Figure 6.2. This is why a crowbar has such a long handle in comparison with its teeth.

If we position the fulcrum so it is nearer to ourselves than to the load, as in Figure 6.4, the lever will convert a large force over a short distance into a small force over a long distance. You might wonder why we would want to convert a large force into a small force. It sounds foolish! But it is not foolish to want to convert a short distance into a long distance. For instance, the oars in a rowboat are levers that convert a short distance into a long distance. The oars are the beams of two levers, and each oarlock is a fulcrum. Since the oarlocks are near the rower's end of the oar, the rower has to use a great deal of force to move the oars. But the rower can make extremely long strokes through the water, even though his arms are relatively short (Figure 6.5).

 Workbook pg. 30

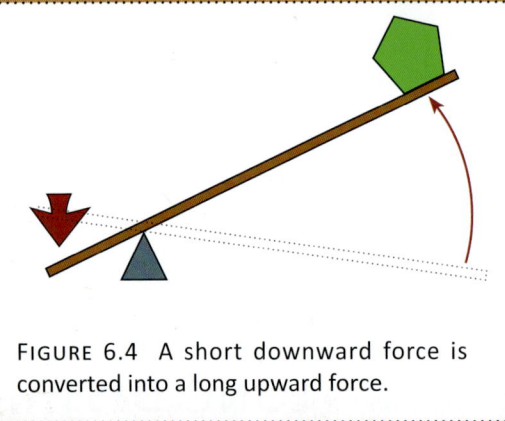

FIGURE 6.4 A short downward force is converted into a long upward force.

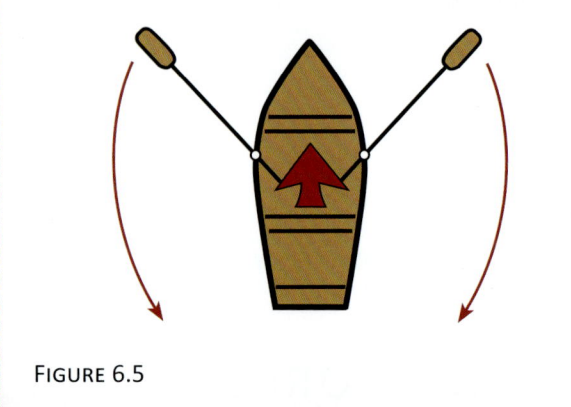

FIGURE 6.5

## 6.6 First, Second, and Third Class Levers

The levers in rowboats, crowbars, and teeter-totters are called **first class levers**. In first class levers, the fulcrum is always between the force and the load.

**First class levers** are levers in which the fulcrum is between the force and the load.

The ancient Greek scientist Archimedes famously declared that if he had a place to stand and a sufficiently long lever, he could even move the Earth.

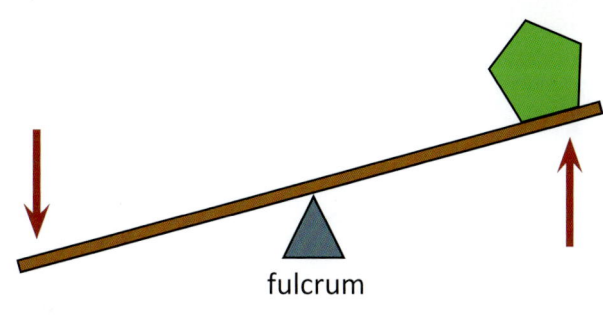

FIGURE 6.6 First class lever

**Second class levers** are levers in which the fulcrum is on one end and the load is between the force and the fulcrum.

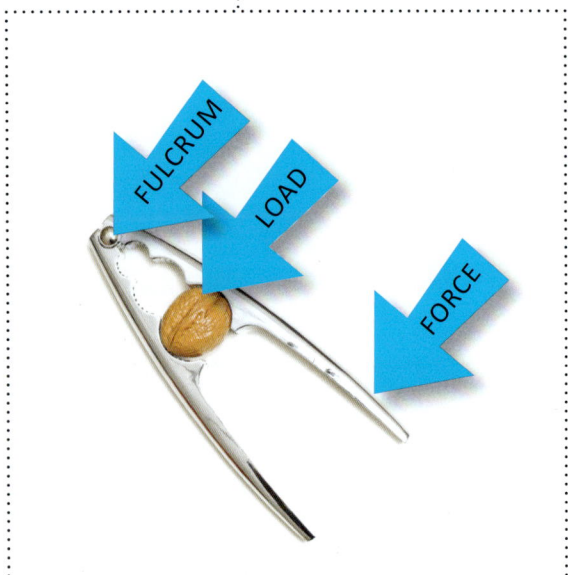

In some levers, the fulcrum is on one end, the force is on the other end, and the load is in the middle. These levers are called second class levers. A nutcracker is a good example of a second class lever. So is a wheelbarrow. **Second class levers** multiply force without changing its direction.

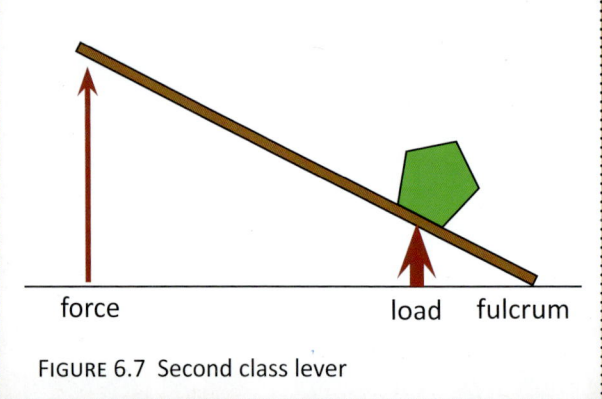

FIGURE 6.7 Second class lever

98

**Third class levers** are levers in which the fulcrum is at one end, the load is at the other, and the force is in the middle.

There is one other type of lever. **Third class levers** are levers in which the fulcrum is at one end, the load is at the other, and the force is in the middle. We use this kind of lever when we sweep the floor. The hand we place at the top of the broom is the fulcrum. The hand in the middle provides the force that moves the broom. The load is the dirt we sweep along the floor, as well as the friction between the broom and the floor.

Third class levers multiply distance instead of force. A tennis racket, a hammer, and a fishing pole are third class levers. So are tweezers and tongs. Third class levers are useful because they multiply distance without changing the direction of the force.

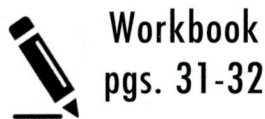

Workbook pgs. 31-32

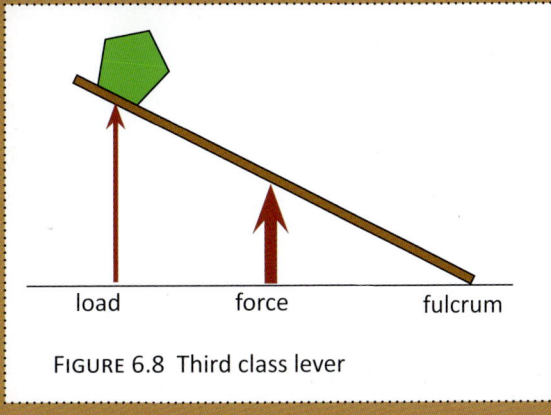

FIGURE 6.8 Third class lever

99

## 6.7 Wheels and Friction

Do you know what friction is? **Friction** is the "drag" that we feel when we move objects across each other. Friction makes it harder to move objects, because they are dragging against other materials. There is a lot of friction between rubber-soled shoes and the ground, so when you walk and run, you have to lift your feet off the ground before moving them forward. But you don't have to lift your feet up to move forward when you ice skate, because there is hardly any friction between your ice skates and the ice.

It takes more energy to move an object if there is a lot of friction between it and other objects. One of the main ways wheels make our work easier is by converting work that would otherwise involve a lot of dragging into work that does not involve dragging.

For instance, if you pull your friend across the yard in a wooden box, there will be a lot of friction between the ground and the bottom of the box. The rubbing between the ground and the box will make your job a lot harder. But if you turn the box into a wagon by adding wheels, there will not be as much rubbing to hold you back. There is still just as much friction between the ground and the wagon—your yard hasn't suddenly turned into an ice-skating rink—but the friction no longer holds you back because you are rolling the wagon instead of dragging it.

A **rolling wheel** makes our work easier because, as the wheel turns, no one part of the wheel touches the ground for more than an instant at a time. As a result, there is little or no rubbing between the wheel and the ground. Wheels are so good at making our work easier that they are used all over the place, in wheelbarrows, wagons, bicycles, strollers, wheel chairs, vacuums, and more.

## 6.8 Wheel and Axle

A wheel can be used to reduce friction, but it can also be used to multiply force the way a lever does. When a wheel is firmly attached to a rod, it becomes a simple machine called the **wheel and axle**.

When you twist a wheel and axle, both the wheel and the axle turn the same amount. If the wheel turns halfway around, the axle turns halfway; if the wheel makes a full turn, the axle makes a full turn. This means that you can do the same amount of work by turning the axle as you can by turning the wheel. The example of a screwdriver makes this easier to understand.

Did you know that a screwdriver is a wheel and axle? The handle of a screwdriver is the wheel, and the neck is the axle. To unscrew a screw by half a turn, you can either twist the handle of the screwdriver halfway around or you can twist the neck of the screwdriver halfway around. The screw turns the same amount in both cases, so the same force is required and the same amount of work is done.

Even though the same amount of work is done in either case, it is clearly easier to turn the handle of a screwdriver than to turn its neck. This is because the distance around the handle of a screwdriver is greater than the distance around its neck. Turning the handle and turning the neck require the same amount of force, but when you turn the handle, your force is spread over a longer distance. This makes your work easier because you don't have to exert as much force at one time.

**Friction** is the "drag" that we feel when we move objects across each other.

A **rolling wheel** converts work that would otherwise involve a lot of dragging into work that does not involve dragging.

The **wheel and axle** is a simple machine made of a wheel firmly attached to a rod.

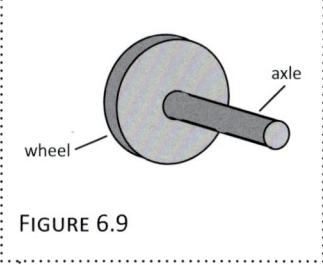

FIGURE 6.9

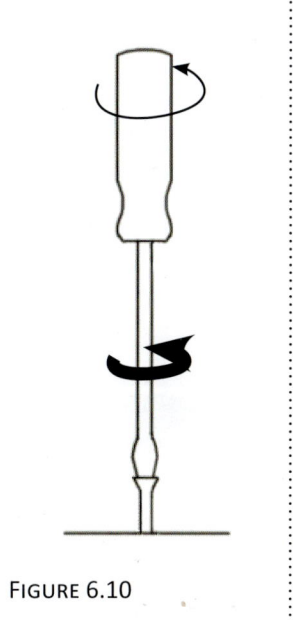

FIGURE 6.10

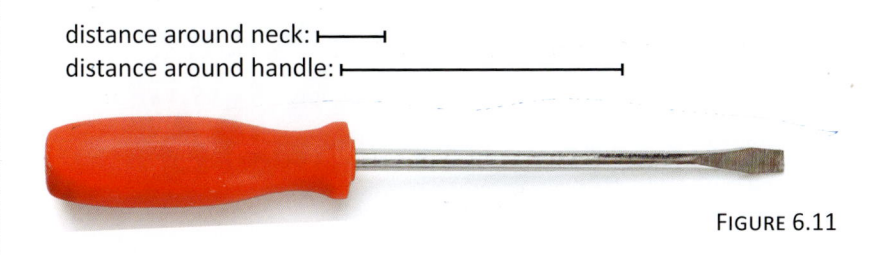

FIGURE 6.11

Steering wheels, doorknobs, faucets, fishing reels, and waterwheels are wheel-and-axle systems like the screwdriver.

There are also reverse wheel-and-axle systems. In a reverse wheel-and-axle system, force is applied to the axle instead of to the wheel. Reverse wheel-and-axle systems are used to convert a small movement into a large movement. For example, a tiny turn of the axle of a Ferris wheel will move the outer edge of the wheel many feet. Other examples of reverse wheel-and-axle systems are an electric fan, the wheels on a car, the propeller on a helicopter, and the beaters on an electric blender.

A reverse wheel and axle multiplies distance, but it does not multiply force. Because the distance around the axle is short, the force must be provided all at once instead of bit by bit. Reverse wheel-and-axle systems can be difficult to turn, so they are often powered by motors instead of by human beings.

Workbook pgs. 33-34

Experiment #19

## 6.9 Pulley

The sixth and last simple machine is the **pulley**. A pulley is a wheel in a frame. When we pass a rope around a pulley, we can use it to lift heavy objects. Much like a lever, a pulley can be used to reverse the direction of a force, multiply a force, or both.

The simplest pulley is a wheel in a frame that is fixed to the ceiling. When you tie a rope to a load and pass the rope over the pulley, you can move the load *upwards* by pulling *down* on the rope (Figure 6.12). This type of pulley, called a **fixed pulley**, changes the direction of your force. Fixed pulleys are especially useful because they allow you to use some of your body weight to lift a load.

Pulleys can also multiply force. When Mike lifts a 20-pound box of books, he has to exert a force of 20 pounds (Figure 6.13). But if he and Daniella carry the box together, they share the weight. She exerts 10 pounds of force and he exerts 10 pounds of force (Figure 6.14).

FIGURE 6.12

A **pulley** is a wheel in a frame.

A **fixed pulley** is attached to the ceiling; it changes the direction of a force.

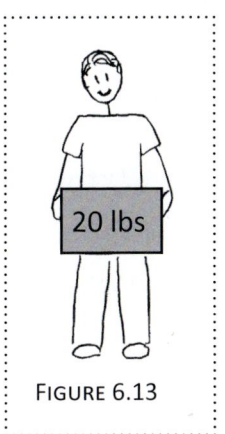

FIGURE 6.13

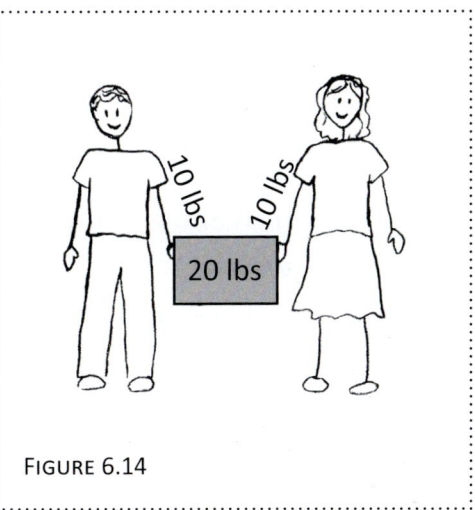

FIGURE 6.14

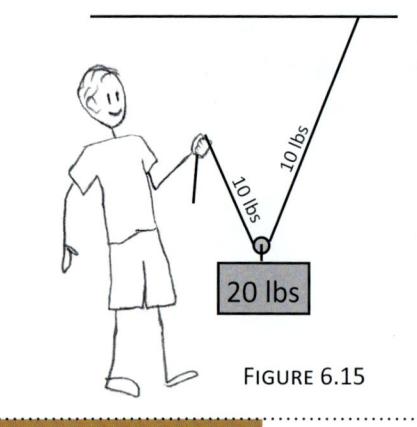

FIGURE 6.15

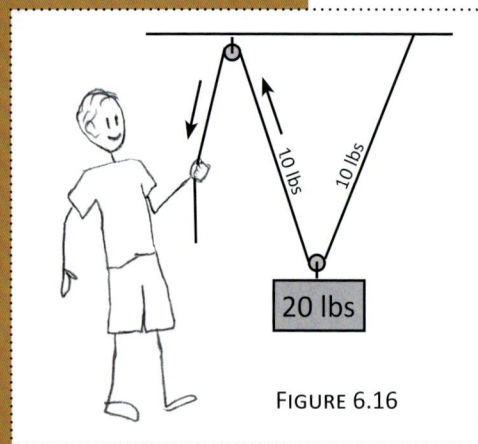

FIGURE 6.16

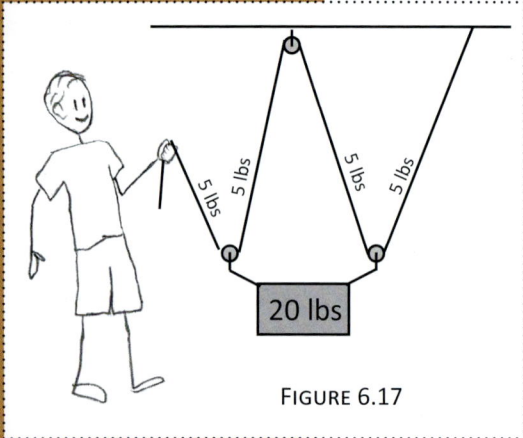

FIGURE 6.17

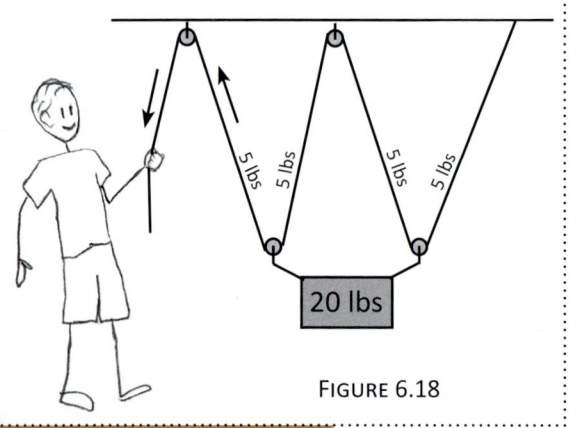

FIGURE 6.18

A **moveable pulley** works in a similar way. A moveable pulley gets its name because the wheel is free to move instead of being fixed to the ceiling. When Mike uses a moveable pulley, he can lift a 20-pound box of books by using only 10 pounds of force (Figure 6.15). Instead of sharing the weight of the load with Daniella, he shares the weight with the ceiling.

Like all other simple machines, moveable pulleys can only multiply force by increasing distance. In Figure 6.15, Mike is lifting the box with only half the force, but he has to pull the rope twice as far. This is because the pulley is supported by two lengths of rope. To raise the box at a distance of one foot, he has to pull until both ropes are one foot shorter, which equals two feet of rope.

A moveable pulley multiplies force, but it does not change the direction of the force. Mike can multiply the force *and* reverse its direction by combining a fixed pulley and a moveable pulley (Figure 6.16). The moveable pulley multiplies Mike's force by transferring some of the weight of the box to the ceiling. The fixed pulley then changes the direction of Mike's force.

Mike can multiply his force even more by using another moveable pulley. When Mike uses two moveable pulleys, as in Figure 6.17, the box's weight is shared by four lengths of rope. Each of these ropes supports a quarter of the weight, but Mike is only holding one of them. So Mike only has to exert a force of five pounds. The rest of the box's weight is held by the ceiling.

FIGURE 6.19

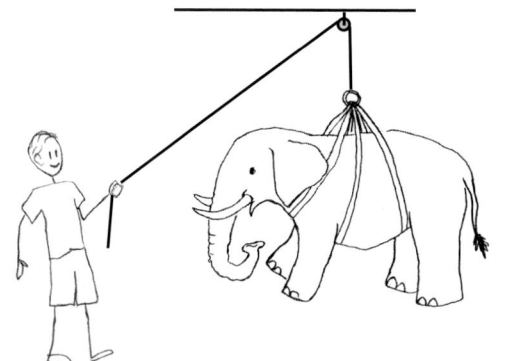

**LARGE FORCE** x short distance      small force x **LONG DISTANCE**

To reverse the direction of the force, Mike can add another fixed pulley at the end of the pulley system, as in Figure 6.18. Since this last pulley is a fixed pulley, not a moveable pulley, it does not multiply Mike's force. It does allow him to pull down instead of up, though.

Notice that Mike is using four lengths of rope in Figures 6.17 and 6.18.[1] This means that, even though he only has to use one-fourth of the force at a time, he has to pull the rope through four times the distance. As always, he can only decrease the force by increasing the distance.

Mike can add as many pulleys as he likes to this system, and each time he adds a moveable pulley, the force he needs to use will decrease. With enough moveable pulleys, he could even lift an elephant! But he would have to pull the rope much, much farther than he would if he used an immense force and no moveable pulleys (Figure 6.19).

A **moveable pulley** is free to move; it reduces the force needed to lift the load.

**Workbook pgs. 35-37**

**Experiment #20**

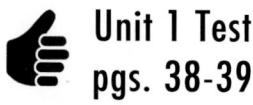
**Unit 1 Test pgs. 38-39**

---

[1] In Figure 6.18, the fourth length of rope has been passed around a fixed pulley, so it might look as if Mike is using five lengths of rope. The rope that Mike is grasping is not directly attached to the box, though, so it does not count as a separate length of rope.

UNIT 2

# 7
Chapter

# Introduction to Biomes

God wills the "interdependence of creatures." The sun and the moon, the cedar and the little flower, the eagle and the sparrow: the spectacle of their countless diversities and inequalities tells us that no creature is self-sufficient. Creatures exist only in dependence on each other, to complete each other, in the service of each other.
— *Catechism of the Catholic Church,* 340

## 7.1 Introduction to Biomes

"Can either of you tell me what a biome is?" Mom asked Mike and Dani.

"It's not a simple machine, is it?" Dani asked with feigned horror. "Mike's been doing so many experiments with machines lately, I was afraid he would end up by taking the van apart."

Mom smiled. "No, it's not a machine. Do you remember the meaning of the Greek root word 'bios'?"

"It means 'life,'" Mike said. "Like *bio*logy. I'm going to be studying life science now," he told Dani.

"That's right," Mom said. "You'll be studying **ecology**, which is the type of biology that explores how the different parts of Creation—plants, animals, weather, terrain—work together. The word 'ecology' comes from the Greek word for house, which makes sense because, in a certain sense, Creation is the house in which the human family lives.

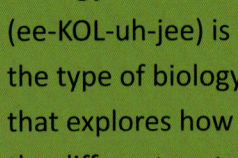

**Ecology** (ee-KOL-uh-jee) is the type of biology that explores how the different parts of Creation—plants, animals, weather, terrain—work together.

"If Creation is like a house," Mom continued, "then biomes are like unique rooms in this house, each possessing its own climate and containing special plants and animals. For instance, look at this picture of a polar bear. See how warm his fur coat is? Do you think he could survive if we moved him to the desert?"

"No," Mike said. "He'd be too hot."

"What about a hummingbird? Hummingbirds have tiny bodies that lose heat quickly, and they use so much energy that they have to eat several times every hour. What would happen if we transported a hummingbird to the arctic tundra, where temperatures are extremely cold and nectar-filled flowers are only available for about one month during the year?"

"The hummingbird would die, too," Dani said.

"That's right," Mom agreed. "But hummingbirds survive and flourish in other parts of the world, such as tropical rainforests. Similarly, polar bears are well-equipped to live in the arctic tundra.

"The rainforest, the desert, and the tundra are three different biomes. **Biomes** are major regions of the globe, each with its own unique climate and its own community of plants and animals. The plants and animals in each biome are perfectly suited to the weather, soil, and terrain of their biome, and many could not survive anywhere else.

"Each biome contains a variety of plant and animal types—called **species**—and these species live in a variety of different **habitats**. A habitat is the area in which an animal lives, and it is different for different species. For example, parrots and peccaries (wild pigs) both live in the rainforest biome, but their habitats are very different: parrots live in the tree tops, and peccaries live on the forest floor.

"Besides having its own habitat, each species in a biome plays a special role, called its **niche**. For example, many flowering plants are pollinated by insects, birds, or bats, so we would say that

**Biomes** (BYE-ohmz) are major regions of the globe, each with its own unique climate and its own community of plants and animals.

**Species** (SPEE-sheez) are types of plants and animals.

A **habitat** is the area in which an animal lives.

A **niche** (NICH) is the special role a species plays in a biome.

When creatures depend on each other for survival, we say they are **interdependent**.

109

A **predatory relationship** is a relationship in which one animal kills another for food.

In **parasitic relationships** (par-uh-SIT-ik), one plant or animal uses another creature for food or shelter, but does not usually kill the other animal.

A **facilitative relationship** is a relationship between creatures in which one or both creatures is benefited and neither is harmed.

pollination is one of the roles, or niches, of these insects, birds, and bats. Similarly, beavers fill an important niche when they cut down trees to construct dams, because the ponds that form behind the dams provide food and homes for dozens of creatures. Birds often fill the niche of seed distribution, and trees fill the niche of providing shelter for birds and small animals. These are just a few examples of how species are **interdependent**."

"What do you mean, 'interdependent'?" Mike asked.

"To be interdependent means to interact with and depend on each other for survival. There are many types of interdependence, but three of the main ways that species are interdependent are through **predatory relationships**, **parasitic relationships**, and **facilitative relationships**," Mom explained.

"Predatory relationships are relationships in which one animal kills another for food. The animal that is eaten is called the prey, and the animal that eats the prey is called the predator. A predatory relationship is good for one creature, but harmful to the other. An example is the relationship between a polar bear and a seal, or between a frog and an insect.

"In parasitic relationships, one plant or animal uses another creature for food or shelter, but does not usually kill the other animal. For example, when a flea sucks blood from a deer, it is not threatening the deer's life. The creature that benefits from a parasitic relationship is called the parasite, and the creature that is harmed is called the host.

"The third type of relationship between creatures is called facilitation," Mom continued. "In a facilitative relationship, at least one animal is helped and neither is harmed. For example, cattle egrets eat the insects stirred up by moving cattle. This relationship is facilitative, because the egrets get a free meal and the cattle are unaffected. The relationship between a dogwood tree and a flock of hungry birds is also facilitative, because neither the tree nor the birds are harmed. In fact, both of them benefit from the relationship, because the birds get to eat the berries on the dogwood tree, and in return they distribute its seeds far and wide in their droppings."

"Is the relationship between flowers and bees an example of facilitation?" Dani asked. "The bee collects nectar from the flower, and it also pollinates the flower."

"That's a great example," Mom agreed.

"I remember you telling us that sometimes a squirrel will bury an acorn and then forget where he put it," Mike said. "So while he works to store up food for himself, he also does a little gardening by planting acorn seeds. Does this count as facilitation?"

"It certainly does," Mom said. "In this case, the squirrel is facilitating the oak tree by planting its seeds, and the oak tree is facilitating the squirrel by providing it with acorns to eat.

"As you can tell," Mom concluded, "facilitative relationships are quite important in Creation, so pay attention during the next few weeks, and see how many you can identify!"

Facilitative Relationship: Cattle egrets eat the insects stirred up by moving cattle.

Parasitic Relationship: Mistletoe (parasite) growing on a tree (host)

Predatory Relationship: Cheetah (predator) with hare (prey)

 **Workbook pg. 41**

A **food chain** is a series of relationships that passes food energy from one creature to another.

A **consumer** is a creature that gets its food energy from other living things instead of by producing it out of non-living materials.

## 7.2 Food Chains

"Facilitative relationships are so neat!" Dani exclaimed. "I like them a lot more than the other relationships between creatures."

"Facilitative relationships are certainly a lot of fun to discover and learn about," Mom agreed, "but predatory and parasitic relationships also show forth the order and intelligence of God's designs. Predatory relationships are often interconnected in complex systems called food chains. For instance, if a rabbit eats grass, and a hawk eats the rabbit, then the hawk is dependent both on the rabbit and on the grass that fed the rabbit. The grass, rabbit, and hawk are connected to each other like the links in a chain, so we say that they are part of the same **food chain**.

"A food chain is a series of relationships that passes food energy from one creature to another. There are at least three types of plants or animals in every food chain: a **primary producer**, a **primary consumer**, and a **top predator**.

"Every food chain begins with a primary producer, which is a creature that converts non-living nutrients and minerals into food energy. Primary producers include plants, algae, and special, deep-sea bacteria. Plants and algae use sunlight to produce food energy through the process of **photosynthesis**; deep-sea bacteria use chemicals to produce food energy through the process of **chemosynthesis**."

Mom paused. "Who do you think the primary producer is in a food chain that consists of grass, a rabbit, and a hawk?"

"Rabbits and hawks aren't plants, and they certainly aren't algae or bacteria," Mike said. "The grass must be the primary producer in this food chain."

112

"That's right," Mom agreed. "In a grass–rabbit–hawk food chain, grass is the primary producer. Unlike the rabbit and the hawk, which get their energy by consuming other living things, the grass produces its own energy out of non-living materials. That's why it is called a primary *producer*.

"The next creature in the food chain," Mom continued, "is called the primary consumer, because it is the first animal in the food chain to *consume* the food energy *produced* by the primary producer. A **consumer** is a creature that gets its food energy from other living things instead of by producing it out of non-living materials. The primary consumer gets its energy by eating the food energy stored in the stems, leaves, and fruit of the primary producer."

"The grass is the primary producer in our food chain, so the rabbit must be the primary consumer," Dani said.

"That's right. The rabbit is the first animal in the chain to consume the food energy produced by the grass. That makes it the primary consumer.

"The last link in the food chain is the top predator, which is the strongest animal in the food chain," Mom continued. "Top predators receive their energy from the prey they kill and eat. Bears, eagles, wolves, and snakes are top predators, but if the other creatures in the food chain are not very large or strong, even a dragonfly could count as a top predator."

The **primary producer** is a creature that converts non-living nutrients and minerals into food energy.

The **primary consumer** gets its energy by eating the food energy stored in the stems, leaves, and fruit of the primary producer.

The **top predator** is the last link and strongest animal in the food chain.

**Photosynthesis** is the process by which plants and algae use sunlight to convert nutrients and minerals into food energy.

**Chemosynthesis** is the process by which deep-sea bacteria use chemicals to convert nutrients and minerals into food energy.

PRIMARY PRODUCER — PRIMARY CONSUMER — TOP PREDATOR

"In our example, the hawk is the top predator, right?" Mike asked.

"That's right. Our food chain is constructed of a primary producer (grass), a primary consumer (rabbit), and a top predator (hawk). Food chains can have more than one consumer, though. For instance, the grass might have been eaten by a caterpillar, which was eaten by a frog, which was eaten by a snake, which was eaten by a hawk. In this case, our food chain would have five links: a primary producer (grass), a primary consumer (caterpillar), a secondary consumer (frog), a tertiary consumer (snake), and a top predator (hawk)."

"Does all the energy in a food chain end up being passed to the top predator?" Dani asked.

"Not all of it," Mom said. "Along the way, much of the food energy is returned to the soil in the form of dead **organic material**. For instance, a hawk cannot digest the skin and bones of a rabbit. Instead of being used as food energy by the hawk, the parts of the rabbit that cannot be digested decompose into nutrients and minerals that enrich the soil. The same thing happens when a plant or animal from any level in the food chain dies. Its body goes back into the Earth as

**Organic material** is or was part of a living creature. Leaves, bones, feathers, and animal droppings (scat) are examples of organic material.

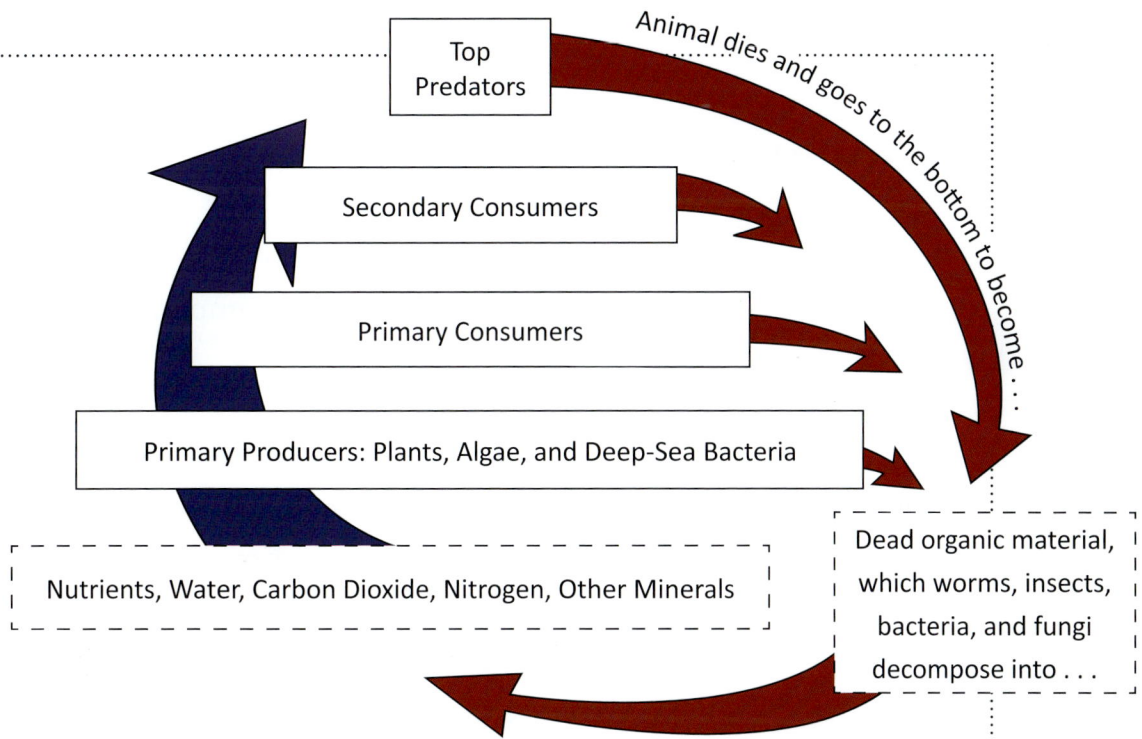

dead organic material, which is decomposed by worms, insects, bacteria, and fungi. The process of **decomposition** enriches the soil by releasing the nutrients and minerals that are stored in organic material. Living plants then absorb these nutrients from the soil to produce food for more plants and animals.

"And since plants are primary producers, they are the beginnings of new food chains!" Mike exclaimed.

"Exactly," Mom smiled. "This is how energy from the Sun and nutrients from the soil are circulated among all the creatures in a biome."

**Decomposition** is the process by which organic material is turned back into soil by decomposers such as worms, insects, bacteria, and fungi.

Workbook pg. 42

## 7.3 Climate Zones Influence Biomes

The Earth can be divided into six different climate zones: the North and South Tropical Zones, the North and South Temperate Zones, and the North and South Polar Zones.

The tropical zones are near the **Equator**, and extend northward to the Tropic of Cancer and southward to the Tropic of Capricorn. The Tropic of Cancer and the Tropic of Capricorn are imaginary, horizontal lines called **latitude lines**. Because of where the Tropic of Cancer and the Tropic of Capricorn are located on the globe, we say that they are the lines at 23.5° north latitude (23.5° N) and 23.5° south latitude (23.5° S).

The North and South Tropical Zones, also called the tropics, include all the parts of the Earth in which the Sun is directly overhead at least once a year. Since the Sun passes so high overhead in the tropics, these zones receive more direct sunlight than do other parts of the globe. As a result, the tropics contain some of the warmest places on Earth, including tropical rainforests and tropical grasslands. The tropics receive the same amount of direct sunlight all year round, so the temperature does not change very

The **Equator** is an imaginary line around the center of the globe, at an equal distance from the North and South Poles.

**Latitude lines** are imaginary, horizontal lines marking a location's distance above or below the Equator. You will learn more about latitude in Chapter 13.

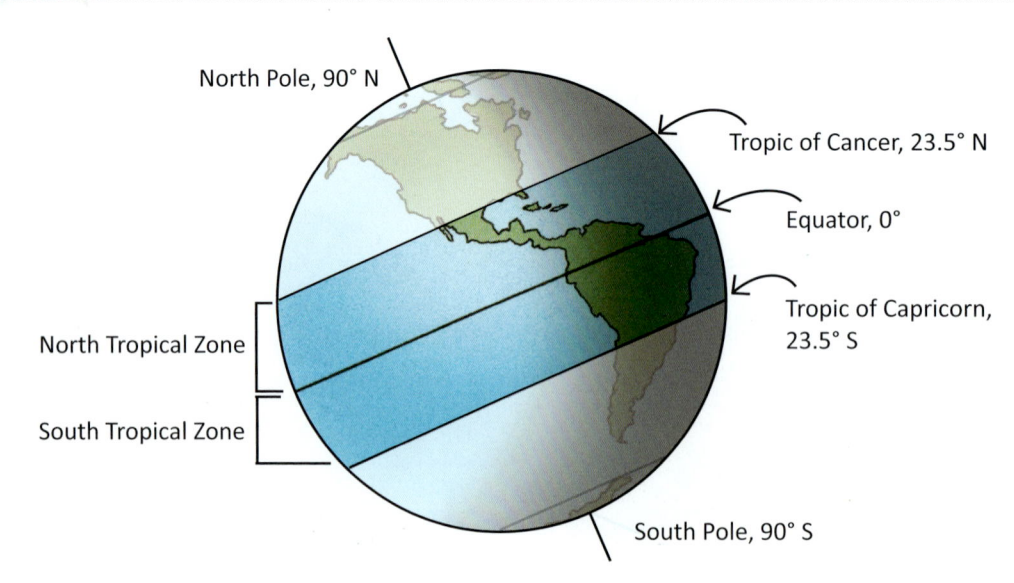

FIGURE 7.1 North and South Tropical Zones

much from summer to winter. The length of day and night also remains the same throughout the year in the tropics.

The North and South Temperate Zones are located in between the Equator and the poles. The North Temperate Zone, containing most of the United States and Canada, extends from the Tropic of Cancer to the Arctic Circle. The South Temperate Zone extends from the Tropic of Capricorn to the Antarctic Circle. Unlike the tropics, the temperate zones receive different amounts of direct sunlight at different times of the year. This is why the temperate climate is sometimes very warm (in the summer) and sometimes very cold (in the winter). The length of day and night also changes throughout the year.

The polar zones are located at the top and bottom of the globe. The North Polar Zone includes everything above the Arctic Circle, and the South Polar Zone includes everything below the Antarctic Circle. Every place in the polar zones experiences at least one day of 24-hour darkness each winter.

Since the polar zones receive little or no sunshine during the winter, they are extremely cold. Even in the summer the sunshine is not very intense, because the Sun remains low in the sky. The

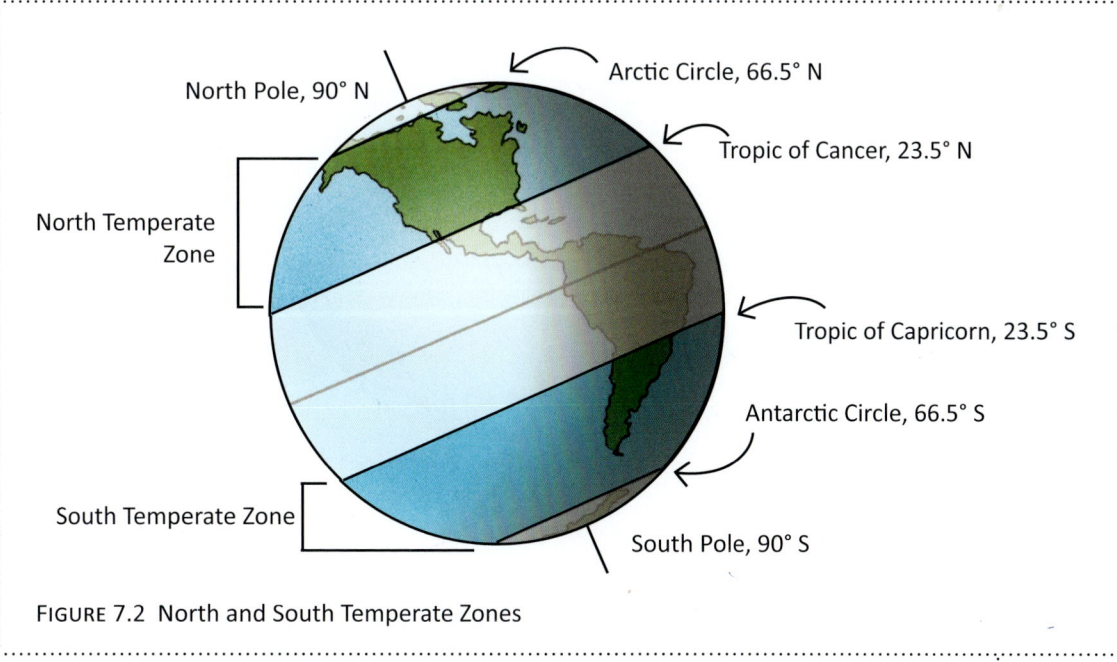

FIGURE 7.2  North and South Temperate Zones

North and South Poles endure the longest periods of 24-hour darkness and receive the least amount of direct sunlight. As a result, they are covered with polar ice caps that remain frozen all year round.

The polar ice caps are not the only type of land in the polar zones. If you travel south from the North Pole, you will eventually leave the polar ice cap and reach a region called the tundra biome. This biome straddles the North Polar Zone and the North Temperate Zone. Here, the weather becomes warm enough in the summer for the snow and ice to melt, which allows for the survival of a greater variety of plant and animal species.

In the next few weeks, you will be learning about seven of the main biomes on Earth: the arctic tundra, the boreal forest, the temperate forest, the tropical rainforest, the temperate grassland, the tropical grassland, and the desert. Since the climate and weather in each biome is greatly influenced by its location on the globe, it will be useful to refer back to the map on the opposite page. So grab your Science Notebook, and let's go exploring!

### South Polar Zone

You may wonder if there is any tundra in the South Polar Zone. The answer is no, because in the places where the weather is warm enough for the snow to melt during the summer, there is only open ocean. This is why most of the creatures that live in Antarctica are marine animals, such as seals. Even animals that spend time on the land get most of their food from the sea: penguins and arctic terns, for example.

**Workbook pgs. 43-44**

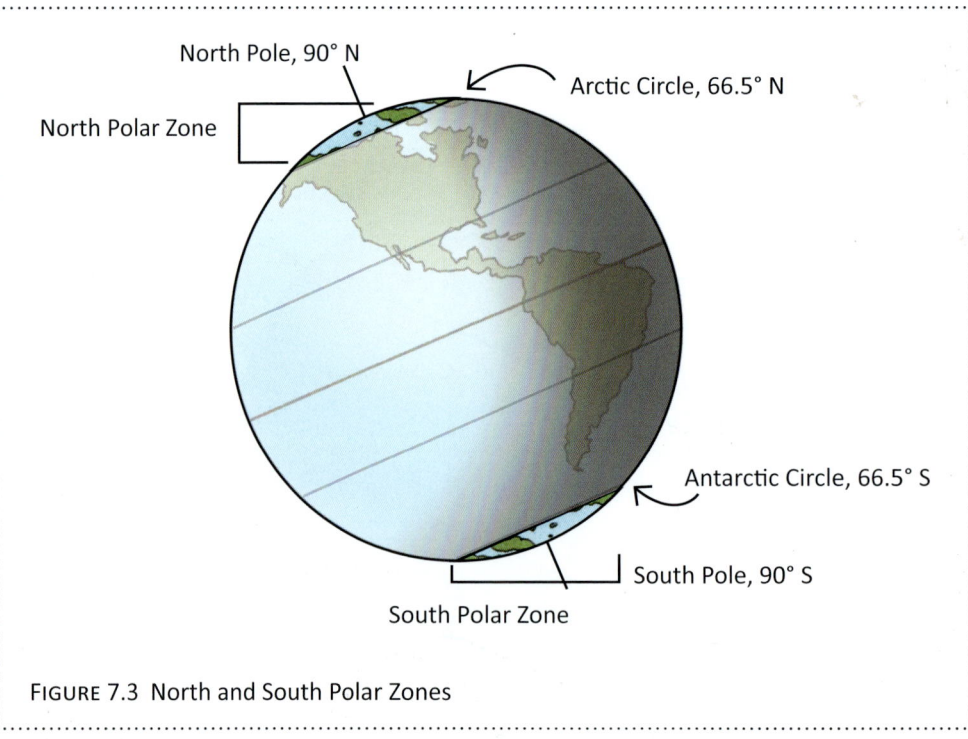

FIGURE 7.3 North and South Polar Zones

# Biomes Map

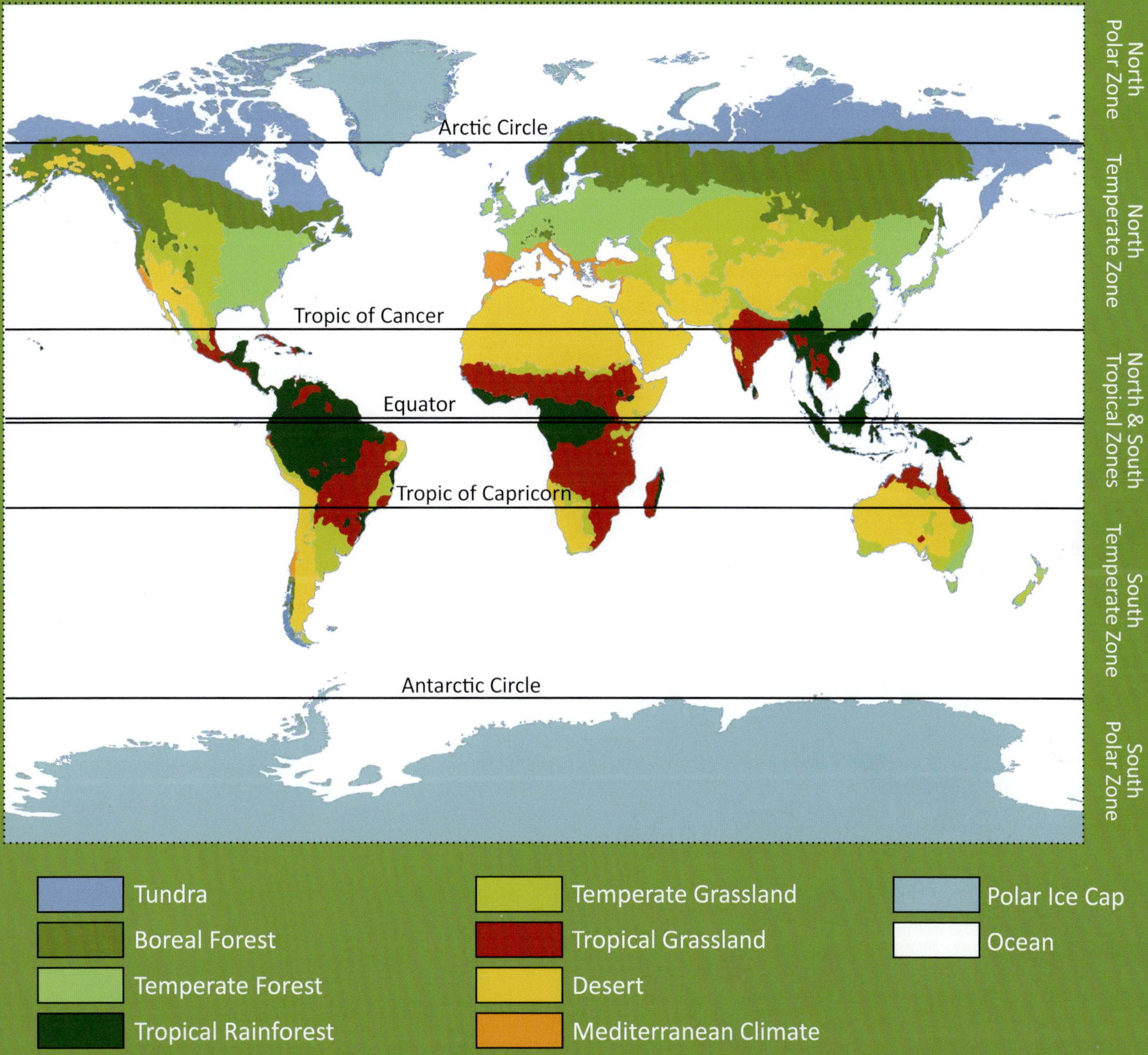

- Tundra
- Boreal Forest
- Temperate Forest
- Tropical Rainforest
- Temperate Grassland
- Tropical Grassland
- Desert
- Mediterranean Climate
- Polar Ice Cap
- Ocean

Map labels: Arctic Circle, Tropic of Cancer, Equator, Tropic of Capricorn, Antarctic Circle

Zones: North Polar Zone, North Temperate Zone, North Tropical Zone, North & South Tropical Zones, South Temperate Zone, South Polar Zone

# 8
## Chapter

# Arctic Tundra

A single swallow, it is said, devours ten millions of insects every year. The supplying of these insects I take to be a signal instance of the Creator's bounty in providing for the lives of His creatures.
— Ambrose Bierce

## 8.1  A Road Trip through the Biomes of the World

Dad cleared his throat and tapped on the breakfast table with the back of his spoon. "I have an announcement to make," he declared, looking at the attentive faces of Dani, Mike, Nick, and Christie. "After much discussion, your mother and I have decided to take you all on a road trip through the different biomes of the world. We'll be visiting the Southwestern deserts, the Amazon rainforest, the African savanna, the arctic tundra, and a few other places, so we have a lot of packing to do."

Nick fell out of his chair in excitement and began doing an Indian war dance, but five-year-old Christie was dismayed by the prospect of visiting the homes of boa constrictors and rattlesnakes. Dani and Mike looked at Dad with disbelief written across their faces.

"Dani," Dad continued, "do you think you can skip your piano lessons for two months, or should we try to bring the piano with us?"

This time Dani caught the twinkle in her father's eyes. "You're just teasing us!" she exclaimed. Mike burst out laughing, and Christie sighed with relief. Nick protested loudly that a road trip was a good idea.

"Of course I'm not teasing," Dad proclaimed with a straight face. "I'm perfectly serious about taking this family on a road trip—a virtual one, that is."

"What's a 'virtuous road trip'?" Christie asked, still a little worried.

Dani giggled. "No, Christie, a *virtual* road trip. Dad means we'll take the road trip in our imaginations," she explained to her younger sister.

"That's right," Dad affirmed. "And considering how active your imaginations are, it should be a pretty exciting trip. We'll 'take off' this evening, after I get back from work. First stop: the arctic tundra!"

"It sounds like we have a lot of packing to do before then," Mom observed. "In fact, let's spend our science hour brainstorming what we should pack. When Dad gets back from work this evening we can discuss our 'suitcase lists' together."

"Agreed," Dad said.

## 8.2 Packing for the Arctic Tundra

After dinner that night, Mike and his family gathered in the living room. "So what have you packed in your suitcases?" Dad asked his family.

"Snow suits!" Nick shouted.

"Gloves and boots and coats and warm blankets," Christie added.

"Snowshoes and skis," Mike suggested. "And sled dogs."

"Don't forget flashlights," Dani added. "The Sun barely rises all winter in the tundra."

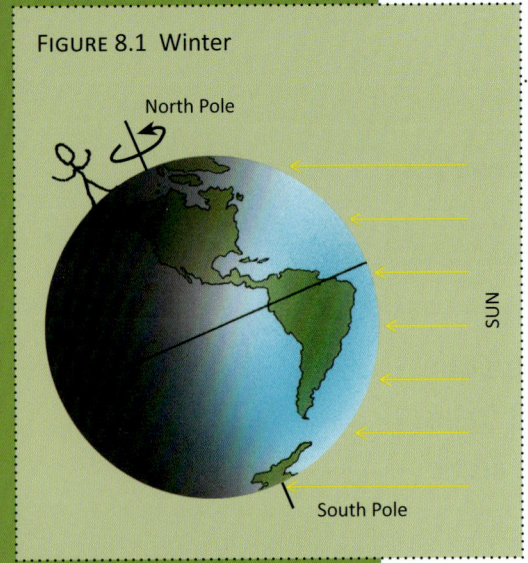

FIGURE 8.1 Winter

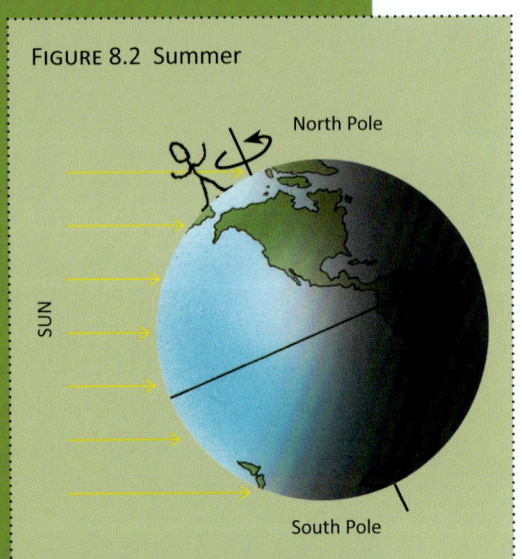

FIGURE 8.2 Summer

As you can see in Figures 8.1 and 8.2, the South Pole is tilted away from the Sun when the North Pole is tilted towards the Sun. This means that when it is summer for the northern half of the globe, it is winter for the southern half of the globe. In Australia, Christmas occurs in the middle of the summer!

"It's dark all day long?" Christie asked in wonder.

"That's right," Dad said. "Who can tell me why?"

"Isn't it because of how the Earth is tilted?" Mike asked. He had learned last year that the seasons are caused by the tilt of the Earth's axis and the revolution of the Earth around the Sun.

"That's right," Dad said. "In the winter, the North Pole is tilted away from the Sun, so it stays in the Earth's shadow all day long. But in the summer, the North Pole is tilted towards the Sun, and then the Sun never sets. It is daytime even in the middle of the night!"

"So I guess we would only need flashlights if we visited the tundra in the winter," Dani said. "In the middle of summer, we wouldn't need them even at night."

"Good point," Dad said. "If we were to visit the tundra in the winter, we would need completely different equipment than if we were to visit it in the summer. For one thing, the winter snow melts in the summer, so we wouldn't need skis, snowshoes, or sleds."

"That explains why my list is so different from everyone else's lists," Mom laughed. "I imagined visiting the tundra in the summer, so I filled my suitcase with lots of bug spray instead of snowshoes."

"Bug spray?" the children asked in surprise.

"Oh, yes! In the summer the air is thick with millions and millions of tiny black flies and mosquitoes. They hatch in the pools of melted snow that cover the ground."

"Ugh!" Dani exclaimed. "I think I'd rather visit the tundra in the winter."

"You'll need to prepare for some pretty cold days and nights, then!" Dad warned. "In the winter the temperature can drop as low as -70°F. At that temperature, perspiration freezes on your body, and your skin will freeze within seconds if it isn't hidden beneath coats and scarfs and gloves."

"Seriously?" Nick asked, awed.

"Seriously," Dad said.

"I'm glad we're only taking a virtual road trip!" Mike said.

## 8.3 Winter and Summer in the Tundra

The tundra is a vast, treeless plain that stretches for miles in every direction. In the winter, the tundra is covered with snow, except where biting winds have blown clear the tops of hummocks and rocks. The average temperature in the winter is -30°F, but it can drop to -70°F or colder. From mid-November to late January, the Sun never rises above the horizon, and the tundra is covered with a shadowy darkness. However, the tundra does not become pitch-black even during the 24-hour nights of winter, because the Sun does not set very far below the horizon.

When spring comes, the Sun begins to rise once more, but only for a few hours at a time. Every day the Sun is visible a few minutes

Depending on how close it is to the North Pole, a particular location in the tundra might experience several weeks of 24-hour darkness in the winter, or only a few days. Locations below the Arctic Circle do not experience 24-hour nights: even in the middle of winter, the Sun rises every day for at least a few minutes.

This Arctic Sun panorama is a 24-hour montage sequence showing the Sun's position in the sky near the horizon for each hour during a single day (June 20-21, 2008) in the arctic summer. Photographed on Axel Heiberg island, in the Canadian Arctic, at 79.5° N.

Dr. Juerg Alean/Photo Researchers

Arctic tundra

longer, until it remains visible all day long. From late May to early August, the Sun traces a wide circle in the sky, just above the horizon.

During this period, the temperatures rise to an average of 40°F. The ice and snow melt, revealing low plants and shrubs that grow among oddly shaped rocks, hummocks, and mounds. Herds of caribou return from the boreal forests south of the tundra where they spent the winter, and millions of migratory birds arrive to take advantage of the abundant insects that hatch in the pools of melted snow that cover the ground.

## 8.4 Permafrost

"The first thing explorers noticed when they visited the tundra was the absence of trees," Dad said. "Can any of my young tundra explorers explain why trees are unable to grow in the tundra?"

"Because it's so cold?" Mike guessed.

**Permafrost** is a layer of ground that stays frozen all year long.

"That has something to do with it, but other plants are able to live in the tundra despite the cold. Why not trees?" Dad paused for dramatic effect. "I can tell you in one word: **permafrost**."

"What's permafrost?" Christie asked. "Is it like frosting on a cake?"

Exposed Permafrost

Dad smiled. "Permafrost is a layer of ground that stays frozen all year long. Because it is so cold in the tundra, the soil never thaws all the way. The first foot or so of soil thaws, but a thick layer of permanently frozen soil, called permafrost, is always present right below the surface. Because the permafrost never thaws, trees are unable to sink deep roots into the soil. This is the main reason the tundra landscape is completely treeless."

"It's also why there are so many insects in the tundra," Mom added. "In the United States, much of our snow soaks into the ground and is taken in by thirsty plants when it melts. But in the tundra, the ground never thaws completely," Mom continued. "Since the melted snow can't soak into the permafrost, it collects in standing pools and marshes. These standing pools are favorite hatching grounds for mosquitoes and other insects. In the summer, the air is filled with clouds of insects."

"In fact, a single caribou can lose nearly a quart of blood each week to mosquito bites," Dad added.

"That's disgusting!" Dani exclaimed.

"I know what you mean," Mom agreed. "But if it weren't for the swarms of mosquitoes in the tundra, thousands of insect-eating birds would starve."

"Wouldn't that make it harder for other animals to find food, too?" Mike asked, remembering what he had learned about food chains.

"Good point!" Dad praised. "If mosquitoes disappeared from the tundra, it would cause starvation all the way up the food chain. If there were no insects, many birds would die, and if there were fewer birds, many of *their* predators, such as arctic foxes and snowy owls, would die, too."

"I guess mosquitoes do some good after all," Dani remarked with surprise.

Mom smiled. "*And God saw everything that He had made, and behold, it was very good*," she quoted. "Everything in God's creation is good if we only have the eyes to see it. Even mosquitoes!"

"I'm still going to slap them if they come near me!" Nick declared.

RUDDY TURNSTONE

PUFFIN

AMERICAN GOLDEN PLOVER

Workbook: Tundra Profile, pg. 45

"Caribou" is the North American name for reindeer.

## 8.5 Arctic Animals

In the winter the tundra is a dark, snow-covered plain. Except for a few polar bears and drifting herds of muskoxen, the windswept tundra seems completely lifeless. But during the short, arctic summer, the tundra bursts into life. Herds of caribou return from their winter feeding grounds, and millions of birds arrive to lay their eggs and raise their young.

FLOCK OF LITTLE AUKS

HERD OF CARIBOU

POLAR BEAR CUBS

Hungry mother polar bears, who have gone without food for many months, emerge from the snow dens in which they gave birth, and begin teaching their roly-poly cubs how to hunt seals. Unless food is in short supply, polar bears eat only the energy-rich blubber of the seals they kill. The rest of the kill is devoured by ravens and arctic foxes, which follow the bears at a respectful distance.

MUSKOXEN

The herds of sturdy muskoxen that roamed the tundra during the winter, nibbling whatever plants were uncovered by the wind, now find more plentiful pasture. If a herd is threatened by wolves, the adult muskoxen form a tight circle around the young calves, guarding them with their mighty horns. Few predators are foolhardy enough to attack a muskox "fortress"!

Rock ptarmigans, which spent the winter camouflaged by their snow-white feathers, prepare to raise broods of young chicks. Every spring they molt, or lose, their white feathers, and grow feathers of a mottled grey and brown. This gives them a better chance to escape the golden eagles, hawks, snowy owls, and foxes that prey upon them and their chicks. The white fur of arctic foxes also changes to brown, helping them avoid predators and sneak up on prey. The search for food is never-ending for tundra animals, who must regain the layer of fat that sustained them during the winter. Arctic foxes collect extra food in the summer and bury it in caches to prepare for the long, cold months ahead.

**Ptarmigan** (TAR-mi-guhn)

The rock ptarmigan's feathers provide it with camouflage both in the winter and in the summer.

ARCTIC LEMMINGS

To one species, the lemming, summer brings increased danger instead of relief. These small, furry rodents look like tailless mice, and during the winter they live in long tunnels between the snow and the ground. There, they nibble on the roots and leaves of snow-covered plants and raise countless offspring. Living beneath the snow, lemmings are close to their food source all winter long.

Although summer brings a renewed growth of vegetation, it also melts the blanket of snow that protected the lemmings from most predators. During the summer, many lemmings are eaten by snowy owls and foxes.

Lemmings are such an important source of food for owls and foxes that the number of owls and foxes each year depends on how large the lemming population is. When there are few lemmings to eat, the owl and fox populations decline; when lemmings are abundant, the owl and fox populations increase, because they have more energy to raise and feed their young. In this way, the populations of predator and prey are kept in balance, maintaining God's good order in the tundra.

SNOWY OWLS

ARCTIC FOX

130

ARCTIC TERNS

## 8.6 Surviving Tundra Winters

"How do you think animals survive the cold, tundra winters?" Dad asked.

"They go to sleep all winter in caves," Christie declared.

"That's a good guess!" Dad said. "In many parts of the world, animals survive cold weather by going into a deep sleep, called **hibernation**.

"Surprisingly, hardly any animals in the tundra hibernate during the winter," Dad explained. "Eight to ten months is just too long for most of them to survive in a state of hibernation. God gave them other ways to survive instead.

"First of all, many creatures avoid the arctic winters by migrating, which means they leave the tundra in the fall and come back in the spring. When winter approaches, vast herds of caribou migrate south into the evergreen forests that border the tundra. Most arctic birds migrate, too. They live in the tundra only during the summer, and spend the winter in a warmer climate."

"Some animals stay in the Arctic all year, right?" Dani asked.

"They sure do," Dad said. "During the dark, arctic winter, many polar bears move over the ice hunting seals for food. Pregnant mother bears, on the other hand, spend the winter in snow dens,

**Migration**

The arctic tern divides its time between the Arctic and the Antarctic, travelling more than 22,000 miles round-trip every year! Because of the tilt of the Earth's axis, the seasons in the Arctic and Antarctic are reversed. When it is winter in the Arctic, it is summer in the Antarctic. The arctic tern enjoys the 24-hour sunshine of the arctic summer, and then migrates south to the 24-hour sunlight of the antarctic summer.

**Hibernation** is a period of winter dormancy. When a plant or animal goes dormant, its body processes slow down so it uses much less energy.

131

Like many other geese, the mother snow goose lines her nest with down feathers plucked from her breast. The down serves to insulate the eggs, while the bare patch on her breast, called a brood patch, helps keep the eggs warm when she is sitting on the nest.

where they give birth to one or two cubs. Whether they spend the winter in the open or in a den, polar bears remain toasty warm all winter, thanks to their dense fur coats. Each of the hairs in a polar bear's coat is a hollow tube containing a little pocket of air. The air inside these tube-like hairs makes the fur such a good insulator that the polar bear actually has more trouble with overheating than with freezing!"

Polar bear hairs, magnified 290x

The children laughed at the thought of overheated polar bears.

"When I was researching polar bears this afternoon, I learned that their fur comes in two layers," Mike said. "The fur closest to the skin is soft, fine, and dense. This undercoat keeps the bear's body heat from escaping into the cold air. Over this first layer of fur is an outer layer made up of long, thick hairs. These hairs form a windbreaker and also make the bear waterproof. Even when he is swimming in the ocean, a polar bear's skin never gets wet."

"Other arctic animals also have two layers of fur," Dad added. "The hairs in a muskox's rough, outer coat are two feet long, and reach almost to the ground! Beneath this outer coat, the muskox is insulated by a coat of soft, fine hairs. When baby muskoxen are born in the spring, they are already covered with a woolly 'sweater' of this fur, and can easily survive -30°F weather."

MUSKOXEN

"I have a question," Mike said. "Mom told us about all the mosquitoes in the tundra, and how important they are as food for birds. But insects don't have fur or feathers. How do they survive the winter? Do they all migrate?"

"I was hoping someone would ask about that," Dad said. "Some arctic insects migrate in the winter, but many use a natural antifreeze to keep from freezing."

Mike and Dani exclaimed in amazement, but Nick asked, "What's antifreeze?"

Some insects do not have to survive the winter; they simply lay their eggs and die. In the spring, their eggs hatch into a new generation of insects.

"Antifreeze is that liquid I poured into the car a few weeks ago. Do you remember watching me? I added it to the engine coolant to keep the coolant from freezing when the weather gets cold." Turning to his older son, Dad added, "Mike can explain to us how antifreeze works."

"I can?" Mike asked, surprised.

"Yes," Dad assured him. "Don't you remember learning that different liquids have different freezing points?"

"Sure! For example, water freezes at 32°F, but other liquids, like rubbing alcohol, won't freeze unless they get much, much colder. But how does this explain antifreeze?"

"Antifreeze is a liquid that has a very low freezing point," Dad explained. "This means it won't freeze in cold weather, and when we add it to engine coolant, the coolant won't freeze either.

"Insect antifreeze works the same way," Dad continued. "**Freeze-avoidant insects** produce chemicals such as glycerol that lower the freezing temperature of their body fluids. The insects' bodies can be as cold as the sub-zero snow and winds around them, but still not freeze!

"There is another type of 'insect antifreeze'," Dad said. "This kind doesn't prevent freezing, but instead allows the insects to survive being frozen. When an animal freezes, ice crystals form in

**Freeze-avoidant insects** produce chemicals that lower the freezing temperature of their body fluids.

and around the cells of its body. Since these crystals have sharp edges, they can tear through the sides of the cell, destroying the cell. An animal usually dies if it is completely frozen.

"This is not the case with **freeze-tolerant insects**! Freeze-tolerant insects produce chemicals that protect their cells by ensuring that any ice crystals form between their cells, not inside their cells. Meanwhile, the cells shrink by becoming partially dehydrated, which gives the ice crystals room to form between the cells without tearing the **cell membranes**. All winter long, the interior of each cell remains in a fluid state, and when summer comes and the insect thaws, none of its cells have been torn by sharp ice crystals."

"It's almost like God knew what He was doing," Dani commented.

Mike objected loudly, "Of course He knew what He was doing!" Then, noticing Dani's grin, he said, "Oh, you're just teasing me."

"No, I was practicing the literary device of understatement," Dani corrected, grinning even more.

Twisting in his chair, Mike attempted to tickle his older sister.

"All right, you two," Dad intervened. "For science tomorrow, I'd like you to research tardigrades. They're also known as water bears or moss piglets. 'Insect antifreeze' is nothing compared with the survival techniques God has given these little animals!"

**Freeze-tolerant insects** produce chemicals that protect their cells by ensuring that any ice crystals form in between their cells, not inside their cells.

The **cell membrane** is the outer layer of a plant or animal cell which holds the cell together.

Workbook: Tundra Research, pg. 45

Workbook: pg. 46

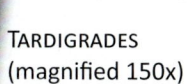
Tardigrades (magnified 150x)

Andrew Syred/Photo Researchers

135

Cotton grass

Bearberry

## 8.7 Tundra Plants in the Winter

Arctic plants have many ways of surviving the winter cold. Some allow all their leaves and stems to die down to the ground, storing a supply of nutrients below the earth in thick roots and bulbs. Other plants are evergreen, and keep their leaves alive all year. This gives them a head start on other plants, since they don't have to produce new leaves every spring. On the other hand, it is difficult for evergreen plants to keep from freezing in the winter.

Surprisingly, the tundra receives very little **precipitation**, only about 5-20 inches per year. In fact, the tundra qualifies as a cold desert, because strictly speaking, a desert is defined as any place that receives less than 15-20 inches of precipitation per year. Unlike hot deserts, the tundra contains many pools of melted snow and standing water in the summer. This is because water does not evaporate quickly at cold temperatures and because the permafrost in the tundra prevents water from soaking very far into the ground. Nevertheless, many plants have to endure periods of drought in the summer, because the water collects on low ground, leaving the hills and hummocks to dry out.

Alpine pasqueflower

**Precipitation** includes rain, snow, and any other moisture a location receives.

Arctic tundra

The arctic willow is related to the weeping willow. Unlike its southern relative, the arctic willow is only a few inches tall.

During the winter, the tundra is covered by a thin layer of snow. The snow acts as an insulating blanket for tundra plants, keeping the heat in their leaves and in the ground from escaping into the even colder air. Although the snow cover can be as deep as several feet, it is often only a few inches deep. As a result, tundra plants are very short and compact, usually only a few inches tall. Snow can keep plants warm because there is a great deal of air trapped between the overlapping snowflakes on the ground. Air is an excellent insulator, so when the temperature outside plummets to a bone-chilling -40°F, a toasty blanket of snow keeps the short arctic plants many degrees warmer.

Many arctic plants provide even more insulation for themselves by growing close together in tightly packed clusters. The air that is trapped between their leaves and stems helps keep them warm. Similarly, many plants hang onto their leaves and stems when they have died instead of shedding them and allowing the wind to carry them away. The dead leaves and stems surround the living plant, acting as an insulating mulch.

Often, the leaves and stems of tundra plants are covered with short "hairs," which serve as insulation all year round.

Experiment #21

Purple saxifrage

The leaves of arctic plants tend to be dark green or red. These dark colors allow them to absorb more of the Sun's warmth.

The **growing season** of a location begins after the last spring frost and ends with the first frost in the autumn.

**Vegetative reproduction** occurs when plants reproduce without seeds.

## 8.8  Tundra Plants in the Summer

"Do you remember getting a letter from your cousins last spring?" Mom asked. "Josh and Hanna were already planting corn, tomatoes, and other vegetables in their garden, while we were still waiting for winter to end."

"Boy, do I remember," Nick sighed. "Christie and I planted snap peas after we got the letter, but they got frozen and turned brown and died."

"That's because we have a shorter growing season than Josh and Hanna do," Mom explained. "Our last frost occurs around the middle of May, and the weather gets cold again towards the end of September. That's a growing season of 120 days, as compared with a growing season of 180 days where your cousins live.

"But our **growing season** isn't short at all compared with the tundra's growing season! In the tundra, the growing season lasts for 60 days or less. That's only two months! Where we live, it is safe to plant a garden towards the end of May, but in the tundra we would have to wait until the end of June or the beginning of July, because the ground would still be frozen. The plants wouldn't have much time to grow, either, because snowstorms and the shorter days of winter begin in the middle of August."

"How are the plants able to survive?" Dani asked in amazement.

"God has given them special 'equipment' so they can grow new leaves, reproduce, and store enough energy for the coming winter in only two months. Tundra plants begin producing energy as soon as the snow melts. Since the Sun barely sets before it rises again, they are able to photosynthesize 24-hours-a-day for much of the summer!

"If tundra plants only had to store energy for the next winter, their job wouldn't be too difficult," Mom continued. "But plants also have to reproduce, so there will be young plants to take their place when they die or are eaten by hungry animals."

Mom paused and turned to her oldest son. "Mike, you learned last year about the different ways plants reproduce. Can you describe them to us?"

Searching his memory, Mike began, "A lot of plants make new plants through flowers and seeds. That's what sunflowers do, and daisies, and snap peas. Sometimes the plants also produce fruit, and the seeds are found inside. Apples and pumpkins are good examples.

"Not all plants need seeds to reproduce, though," Mike continued. "For instance, strawberry plants send out runners along the ground with baby strawberry plants at the end. Bulbs like irises and onions reproduce by dividing their bulbs into two or three plants. And the 'eyes' on a potato will sprout as new potato plants if they are left in the ground."

"Or in the closet!" Nick grinned, remembering the "potato monster" he had made by placing a potato in a paper bag and storing it in a dark closet for a month.

"Great description, Mike," Mom said. "When plants reproduce without seeds, it is called **vegetative reproduction**. Many of the plants in the tundra reproduce by vegetative reproduction, because there aren't always enough pollinating insects for all the flowers. Bees are the main pollinating insects, and even though there are many mosquitoes and flies in the Arctic, there are very few bees."

"What is 'pollinating'?" Christie asked, looking up from the polar bear coloring page that she was covering in glue and fuzzy cotton balls.

"Pollination happens when pollen from one flower is carried to another flower," Mom explained. "This allows the flower to form seeds. If a flower is not pollinated, it will die without making seeds.

"Tundra plants that don't reproduce by vegetative reproduction," Mom continued, "are sometimes pollinated by the wind or are self-pollinating, but many still depend on insects for pollination. God has given these plants special 'equipment' to attract insects."

"Equipment?" Nick asked. "Like the backpack and water bottle I brought on my hike with Dad?"

"Not exactly," Mom smiled. "But some tundra flowers do carry heaters with them."

"No way!" Mike exclaimed.

Figure 8.3

Mountain Avens

140

"Yes," Mom affirmed. "The petals of some flowers, such as the mountain avens, have a special shape that allows them to act as heaters for insect pollinators. This shape, called a **parabola**, concentrates the Sun's rays onto the center of the flower. The warm spot in the center of the flower encourages insects to pollinate the flower, and gives them a boost of warmth and energy before they move on to pollinate other flowers."

"It probably helps the flower stay warm, too," Dani remarked.

"Good point. Keeping their flowers and seeds warm is so important that some tundra flowers actually move to catch the sunlight. The flowers of the arctic poppy turn to face the Sun all day long, just like sunflowers do."

"Except that in the tundra, a summer day is 24 hours long!" Mike added. "No wonder arctic plants are able to get so much done in such a short growing season!"

ARCTIC POPPY

"God has certainly equipped them well for survival," Mom agreed. "In fact, He has even made it so that some plants are able to reproduce over the course of several summers. For instance, a plant might bloom one year, and form fruit the next year. In the third year, the fruit ripens, and the seeds are dispersed."

"So in the tundra, we'd have to wait three years to get apples off our apple tree," Nick concluded mournfully.

A **parabola** (puh-RAB-uh-luh) looks like half of a circle or oval.

Since their life cycle proceeds so slowly, arctic plants can live for hundreds of years, growing only a tiny amount each year.

## 8.9 Lichen

**Lichen** (LYE-kuhn) is one of the main foods of caribou, and is formed from two different creatures: fungi and algae.

**Fungi** (FUN-gye) are plant-like creatures that get their energy by decomposing organic matter.

**Algae** (AL-jee) are tiny creatures that produce their energy through photosynthesis, and have a less complicated structure than plants.

The most important "plant" in the tundra is not a plant at all. It is called **lichen**, and it is one of the main foods of caribou. All summer long, caribou graze on grasses and lichens in the tundra. When winter arrives, the caribou migrate south to the boreal forest, where lichen is also plentiful. Foraging for food, the caribou use their large, broad hooves to dig for lichen beneath the snow. Most animals are unable to digest lichen, but caribou have special bacteria in their stomachs that help them break down the complex carbohydrates and nutrients in lichen.

CARIBOU

Lichen is formed from two different creatures: fungi and algae. Mushrooms, mold, and yeast are types of fungi you might be familiar with. You have probably seen algae before, too, perhaps clinging to a rock in a pond or creek.

**Fungi** are not able to produce their own energy through photosynthesis. Instead, fungi are decomposers and get their energy by absorbing the nutrients in organic matter, such as old tree stumps and piles of dead leaves. Fungi are important members of almost every biome because their work of decomposition releases the nutrients in dead organic materials back into the soil.

**Algae** also play an important role in nature. Like trees and grass, algae are primary producers because they are excellent photosynthesizers. The "green slime" in your neighbor's pond is busy turning sunlight into food energy, which is eaten by minnows, which are eaten by larger fish, which are eaten by—you!

Very few, if any, fungi and algae could survive in the tundra by themselves. But when algae and a fungus join together to form lichen, they become one of the hardiest members of the tundra biome. Unlike ordinary fungi, lichens do not have to absorb their energy from dead organic matter, because the algae in the lichen produces energy through photosynthesis. In return, the fungus absorbs water and minerals for the algae to use and provides the algae with a secure place to live. The fungus part of the lichen absorbs the water in the air through their "leaves," which means that lichens can grow even on bare rock.

Lichens are particularly well-equipped to survive drought in the tundra, because they are able to dry out completely without suffering harm. When water is not available, lichens simply stop growing, shrivel up, and wait for the next rainfall or snowmelt. Because lichens are dormant much of the time, they grow very slowly—in fact, some lichens in the tundra might be more than one thousand years old.

Even though they are able to survive extreme cold and drought, lichens are very fragile. As with many other tundra plants, even stepping on them can kill them.

 **Workbook pgs. 47-49**

 **Experiment #22**

Which is the primary producer in this food chain? Which is the top predator?

# Chapter 9

# Boreal Forest

To me a lush carpet of pine needles or spongy grass is more welcome than the most luxurious Persian rug.
— Helen Keller

## 9.1 Boreal Forest

"Is everyone ready to explore another biome?" Dad asked. His young explorers cheered in excitement.

"Then let the virtual road trip continue! We've spent the last two weeks in the arctic tundra, near the top of the globe. Now we're going to travel south to the Canadian **boreal forest**, following the migration routes of the caribou."

"Do we get to pack our suitcases again?" Christie asked. When they had visited the tundra, she had enjoyed drawing pictures of coats, snow boots, and bug spray to fill the "suitcase" that Dani outlined for her on a piece of drawing paper.

"Let's do that right now," Dad said. "What do you think we should bring?"

The **boreal** (BOR-ee-uhl) **forest** is also known as the **taiga** (TYE-guh).

"The boreal forest is still pretty cold, right?" Dani asked. "We should pack warm clothes."

"Definitely," Mom agreed. "The boreal forest gets its name from the Latin word 'boreus,' which means 'northern.' The boreal forest isn't as close to the North Pole as the tundra, so it doesn't have 24-hour nights in the winter. But sometimes the days are only four hours long."

"Then I'm gonna bring my flashlight again!" Nick declared.

"Of course, in the summer the days are 20 hours long," Mom added. Christie looked pleased, and Dani voted to visit the boreal forest in the summer.

"Since it's the boreal *forest*," Mike said, "I'll bring an axe. That way I can build a log cabin for us all to live in."

"A fishing pole and rifle would also be useful," Mom told him. "The lakes and forests of the boreal forest are full of fish and wildlife."

"I'd advise you to bring rubber boots," Dad said. "There are hundreds of bogs and fens in the boreal forest. These can look solid, but they are actually thick mats of soggy moss and grass that float on top of large pools of water. You might start walking across one, and then suddenly sink through the spongy grass and moss into dirty water up to your knees, or even your waist!"

"Ugh!" Dani shuddered.

"These bogs are actually quite fascinating," Dad assured her, "especially the meat-eating plants."

"Meat-eating plants?!" Nick exclaimed. "Let's go!"

## 9.2 Weather and Soil

The boreal forest is the largest of all land biomes, covering thousands of square miles in North America, Europe, and Asia. The boreal forest is a patchwork of forests, lakes, and wetlands. These northern forests consist mainly of **evergreen**, coniferous trees such as pine, fir, cedar, and spruce. Tamarack, or larch, a **deciduous conifer**, is also common in the boreal forest.

Like the tundra, the boreal forest is bitterly cold in the winter, with temperatures ranging from 30°F to -50°F. The creatures of

**Evergreen** trees are trees that keep their leaves through the winter.

**Deciduous** trees are trees that shed their leaves every winter.

**Conifers** are trees that produce cones.

A **bog**, or **muskeg**, is a type of wetland that forms over standing water. Bogs are dominated by sphagnum moss, which forms a thick layer of peat.

A **fen** is a type of wetland that receives water and nutrients from surrounding streams or groundwater. Fens contain a greater variety of plants than bogs do.

the boreal forest must also adapt to comparatively hot weather in the summer, when temperatures can range from 70°F to 100°F. The boreal forest in eastern Russia has experienced lows of -90°F in the winter and highs of 90°F in the summer! Since boreal forest summers are longer and warmer than tundra summers, there is little or no permafrost. Although the ground freezes in the winter, it does not remain frozen through the summer. Because the ground is not frozen solid all year as permafrost, trees are able to grow in the boreal forest, and these trees provide a habitat for a wide variety of forest animals.

Besides extreme temperatures, plants in the boreal forest must also endure long periods without water. Some parts of the boreal forest receive as much as 40 inches of precipitation per year, but other regions receive as little as 15 inches. Over half of the precipitation in the boreal forest occurs in midsummer, and the rest occurs as snow during the winter. Snowfall in the boreal forest varies greatly from place to place; in some areas the snowfall is only one or two feet, while in others the snow cover might be 10-12 feet deep. Ten to twelve feet of *snow* is not the same as 10-12 feet of *precipitation*. Snow, especially in the boreal forest, is so fluffy that several feet of snow might be equal to only an inch or so of liquid water.

Compared to the nutrient-rich soil of southern forests, the soil of the boreal forest is quite poor. The acidic needles of conifer trees require a long time to decompose, so instead of turning into a rich humus, they form a thick, spongy layer on the forest floor. The cold temperatures during much of the year delay decomposition even further, because it is difficult for worms, insects, fungi, and bacteria to decompose organic material without warm weather. Nevertheless, the boreal forest is full of vegetation, especially evergreen trees, water plants, shrubs, mosses, lichen, and fungi. All of these plants[1] are experts at surviving drought and obtaining nutrients.

The boreal forest also contains many wetlands. One type of wetland is called a **bog**, or **muskeg**. A bog is a soggy, flooded area dominated by sphagnum moss. The sphagnum moss floats on top of standing water and forms a thick, spongy mat called peat. Growing thicker each year, this soggy mat of sphagnum moss often becomes solid enough to walk on, and can even support the growth of shallow-rooted bushes and trees, especially around the edges of the bog. The soil is exceptionally acidic and nutrient-poor, so it is difficult for most plants to thrive in a bog.

**Fens** are another type of wetland in the boreal forest. Whereas a bog forms over standing water, a fen receives water from surrounding streams or groundwater. This provides the fen with a slow supply of nutrients. A fen is much like a bog in appearances, except that it contains a greater variety of plants, including tough water grasses, reeds, and shrubs. Fens are much more treacherous than bogs. The islands of floating moss and grass in a fen often seem solid, but can give way without warning beneath the feet of unwary travellers.

Workbook: Boreal Forest Profile, pg. 50

---

[1] Strictly speaking, of course, lichens and fungi are not plants.

Moose

Grizzly bear

Caribou

Lynx pursuing snowshoe hare

## 9.3 Boreal Forest Animals

Despite the hardships imposed by its extreme climate, the boreal forest is home to an abundance of animal life. In the summer, the trees are full of chattering red squirrels jumping from branch to branch. Red crossbills pull seeds from pine cones, and woodpeckers hammer away at trees in pursuit of insects. The secretive fisher moves fluidly through the treetops in pursuit of birds, or hunts hares and lemmings on the forest floor. Long-legged moose browse on willows and poplars, and wade into lakes to eat water plants. Grizzly bears amble slowly as they feast on the abundant blueberries.

In the fall, thousands of migrating caribou arrive from the northern tundra to spend the winter. When winter arrives with its deep blanket of snow, moose and caribou plough through the snow with their long legs and heavy bodies, searching for edible mosses, lichens, and twigs. The trails made by these huge animals are used by smaller animals, such as foxes and wolves, as they travel through the forest in pursuit of their next meal. Some animals, such as the snowshoe hare and the lynx, are able to travel without sinking into the snow. Their large, furry feet act as snowshoes and permit them to run on top of the snow.

During the long, dark winter, freezing temperatures and biting winds pose a challenge for every boreal forest animal. Caribou shelter behind large bushes, while birds and small animals huddle in the branches of evergreen trees. There, the leafy branches provide them with warm little shelters, out of the wind and insulated from the freezing temperatures of the boreal winter.

The twisted beak of the red crossbill is perfect for extracting seeds from pine cones. Legend says that the crossbill's beak was twisted on Good Friday, when it tried to pull the nails from Our Lord's hands and feet. Its feathers are red, as if stained by Christ's blood.

The fisher is an agile member of the weasel family. Despite their name, fishers eat just about everything but fish, including porcupines! One of the few predators of the porcupine, the fisher attacks the porcupine's face to avoid being stuck with quills.

Unlike the teeth of most animals, a beaver's teeth are covered with a hard, orange enamel. This enamel contains a great deal of iron, which explains both the strength and color of the enamel.

## 9.4 Beavers

"Who wants to share what they've discovered today about the boreal forest?" Dad invited.

"I researched beavers this afternoon," Dani began. "Did you know they are in the same family as mice and squirrels? They are the largest North American rodent and the second largest rodent in the world!"

"I didn't know they were rodents," Mike said. "That explains why they have such big front teeth, like mice and hamsters!"

"Yes, I guess it does," Dani agreed. "Beavers use their teeth to chew through wood. Their jaws are so strong that it only takes them a few minutes to cut down a small tree. A large tree might take several days, though."

"Do they eat trees?" Nick asked.

"Not exactly. They eat the twigs and small branches of the trees they cut down. If the tree is young and tender, they will also eat its bark. Then they use the rest of the tree to build dams and houses."

"They build houses?" Christie asked with interest.

"Yes," Dani replied. "They build big, dome-shaped houses in the water, called lodges. The lodges are made of logs and twigs and mud that the beavers pack together. In the winter the mud freezes as hard as concrete! The lodges can be 10 feet tall or more, although only about five or six feet will show above the water."

"Why don't they drown if their houses are underwater?" Mike asked.

"The entrance to the lodge is deep underwater, but the top is above the water level. To make a lodge, beavers first pile up a huge mound of sticks and mud. Then they dig and chew a tunnel in the mound, beginning underwater and moving towards the top of the mound. When their tunnel rises above the water level, they expand it into a large chamber, which stays dry and warm all winter.

"At least, it stays warm compared to the temperature outside," Dani amended. Looking at her notes, she explained, "The inside

Beavers communicate with other beavers by slapping their tails on the surface of the water.

of a beaver lodge stays at around 32°F even when the temperature outside drops to -50°F. This is how the beavers survive during the winter."

"Where do they get their food in the winter?" Dad asked.

"They work all summer to store up a huge pile of branches underwater near their lodge," Dani explained. "Then in the winter, they swim under the ice from their lodge to their store of food, and eat the twigs and small branches.

"They also eat a lot during the summer and store the extra energy as fat. Guess where the beaver stores most of its fat!"

"Around its tummy?" Nick guessed.

"No, in its tail!" Dani exclaimed. "It has a wide, flat tail that it uses as a rudder when it's swimming, and during the summer the tail gets really thick and fat. The beaver uses the fat in its tail during the winter."

"I didn't know beavers used their tails as rudders," Dad remarked. "It sounds like they are perfectly designed to live in the water."

"Definitely," Dani agreed. "They also have webbed hind feet to help them swim, and their nose, throat, and ears close when they are underwater. Beavers even have transparent eyelids, like goggles!"

"I read once that when a male beaver and a female beaver mate, they stay together for life," Mom remarked. "Do you know if this is true?"

"Yes," Dani said. "And a beaver colony belongs to a single family only. The family might include two adult beavers, four or five yearlings, and a litter of beaver kits. The beaver family has its own territory and will fight off other beavers."

"You said that beavers built lodges, but don't they build dams, too?" Mike asked.

"They sure do! Beavers need deep water to survive the winter," Dani explained. "If the water isn't deep enough, their underwater supply of food becomes covered in ice, and the beavers starve. Besides, it is easier for them to move around in water than on land. So when beavers can't find a pond, they make their own! They use the trees they've chewed down to build dams across creeks and streams. They make the dam thick and sturdy with branches, rocks, and mud, and eventually, they end up with a large, deep pond to make a home in."

"That was a fascinating presentation, Dani," Dad complimented her. "Beavers sound like amazing animals, and marvellous examples of how God equips His creatures for survival."

"Beavers help other animals to survive, too," Dani added. "The ponds they make provide a habitat for fish, frogs, and insects. Then these become a source of food for birds and otters."

"I always find it amazing how God's creatures work together," Mom said. "To be sure, there is a lot of competition in the animal world, but the different creatures also depend on each other. It's like a foreshadowing, or hint, of how God made human beings to help each other."

"Especially with chores," piped up Christie. "Do you think baby beavers have chores to do?"

**Workbook:
Boreal Forest Research Assignment, pg. 50**

**Workbook, pg. 51**

**Experiment #23**

## 9.5 Evergreen Forests

Although the boreal forest biome contains bogs and fens, lakes and rivers, it is first and foremost a conifer forest. Millions of pine, cedar, and fir trees grow for miles in every direction. Young saplings and shrubs grow in the dim light beneath the branches of the larger trees. Mosses and mushrooms multiply on the forest floor.

Not every kind of tree can grow in the boreal forest, but God equipped conifers with all they need to survive there. First of all, the pointed shape of conifers helps keep their branches from cracking under the weight of a heavy snowfall. Instead of pointing upward, the branches of conifers can flex towards the ground, which helps the snow slide off the tree.

Most conifers are evergreen, which means that their leaves stay green all year round, unlike the leaves of deciduous trees. Instead of dying and falling off all at once in the autumn, like the leaves of deciduous trees, the needles of evergreen trees fall off a few at a time throughout the year. Evergreen trees don't have to produce a whole new set of leaves each spring, so they don't require as many nutrients as deciduous trees. This is one reason why evergreen trees are able to grow in the nutrient-poor soil of the boreal forest.

There is another benefit to being an evergreen in the boreal forest. When the days grow longer in the spring, evergreen trees are able to start photosynthesizing immediately, instead of having to grow new leaves first. Since the summer is so short in the boreal forest, this is an important advantage. The dark green color of most evergreens is also helpful, because it allows them to absorb more of the Sun's warmth.

Have you ever wondered why a pine tree has needles instead of large, broad leaves? The reason is that needles are better at conserving, or saving, water than broad leaves are. Winter is like one long drought to the evergreen trees in the boreal forest. The ground might be covered with snow, but until it melts in the spring, the trees will not be able to use it. The trees must survive the entire winter using the water they have stored in their roots and branches. Needles help evergreen trees survive the winter because water evaporates quickly from broad leaves, but slowly from needle-shaped leaves. Plus, needles are usually covered with a waxy coating, which helps them to conserve water even better.

The first chewing gum was lumps of pitch from trees!

## 9.6 Forest Fires

God made only one mistake when He created conifers—or at least, it might look like a mistake at first. The branches and needles of conifers are full of sticky **resin**, or pitch. This resin makes conifers catch fire more easily than other trees, so there are many more forest fires in the boreal forest than in other forests. The dry mosses and lichens that cover the ground and branches of boreal forest trees are also highly flammable, especially after a period of drought.

A conifer will be busy conserving water in the winter and producing cones in the summer. Suddenly, it is struck by lightning during a thunderstorm, and bursts into flames! The fire spreads rapidly from resin-filled tree to resin-filled tree, or through the dry pine needles, mosses, and lichens on the forest floor. After a severe fire, all that is left are ashes and charred tree trunks.

Wait a minute! Can God ever make a mistake? Do you remember

**Resin** (REH-zin)

how difficult it is for dead leaves and branches in the boreal forest to decompose? It is difficult for plants to get the nutrients they need in the boreal forest because it takes a long time for dead organic materials to decompose into nutritious humus.

But when fire sweeps through the forest, the spongy carpet of dead conifer needles, branches, cones, and other organic material is decomposed in just a few minutes by being converted into ash. These ashes release nutrients back into the soil, making the ground rich and fertile. New plants can grow up strong and healthy on the burned ground.

Forest fires also clear the ground for new growth and ensure that new seedlings have plenty of light. Instead of growing in the shade of mature trees, the seedlings can sprout in bright sunshine. Clearly, God did not make a mistake when He created conifers to be full of resin, and filled boreal forests with flammable moss and lichen! He knew all along that frequent forest fires would help, not harm, the boreal forest.

JACK PINE

When fire sweeps through the boreal forest, some birds and animals die, of course, but many escape by flying or running away. Even many slow-moving animals escape by burrowing into the ground. As soon as the burned area cools, the forest inhabitants begin to return, often beginning with bark beetles that lay their eggs on the dead trees.

The jack pine is a good example of God's perfect planning. This type of tree depends on forest fires to produce new jack pines. Most pine cones release their seeds in the fall, but jack pine cones are sealed shut with resin. They only open when they are exposed to the heat of a fire.

Forest fires destroy mature jack pines, but they also release the seeds that will grow into new jack pines. When the fire has passed, thousands of seeds sprout in the burned ground. Jack pine seeds grow best in soil that has been recently burned, so they quickly become young saplings, stretching their branches in the sunshine.

Jack pines mature faster than almost any other conifer in the boreal forest, so the burned area is soon a forest again. Within two or three decades, shade-loving trees begin to grow beneath the jack pines, and pine needles, moss, and lichen cover the branches and forest floor. Before a hundred years have passed, another fire will begin the process all over again with nutritious ash, bright sunlight, and new jack pine seeds.

Frequent fires also purify the boreal forest of diseases and parasitic insects. For instance, jack pine budworms are caterpillars that eat the needles of jack pines. When a stand of jack pines is infested with jack pine budworms, fire can benefit the forest by killing the budworms. New jack pines can then sprout and grow up to be healthy trees.

At first, scientists thought fires were harmful to forests. How could a destructive fire be good for a forest when all it leaves behind is ashes and dead trees? Now they understand that some forests need forest fires to remain healthy. The burned area might seem ugly and lifeless for a while, but it will soon be full of new life. In God's perfect plan for the boreal forest, even forest fires work for the ultimate good of the plants and animals in the biome.

God's plans for our own lives can also be hard to understand. Sometimes things happen that seem just as destructive as a forest fire! But the example of the boreal forest teaches us that we can always trust our loving Father. "I know well the plans I have in mind for you," God says in Jeremiah 29:11, "plans for your welfare and not for woe, so as to give you a future of hope." No matter what happens in our lives, God can use it to bring us closer to Him, filling our souls with new life. We can trust God's loving plans!

Forest fires are an integral part of the boreal forest biome, and scientists estimate that most areas of the boreal forest burn regularly every 50–200 years. This is quite often compared to the infrequency of fires in other forests.

 Experiment #24

PITCHER PLANT

BLADDERWORT

## 9.7 Carnivorous Plants

"When do we get to learn about meat-eating plants?" Nick asked eagerly.

"I think Mike can tell you about them," Mom said. "He spent his science hour this afternoon researching carnivorous plants. Carnivorous means meat-eating," she explained to Nick and Christie.

"Carnivorous plants capture and eat insects," Mike began. "God gave them special shapes so they can capture the insects they need for food. Sundews grow sticky tentacles so that when an insect lands on them, it can't get free. Bladderworts suck insects into little, underwater traps.

"My favorite carnivorous plant is the pitcher plant," Mike continued. "It is shaped like a pitcher or a cup of water, and it is partially filled with rainwater. The plant produces nectar to attract insects, and when they crawl down into the pitcher, they drop into the water and can't climb out again. Then the plant sends out juices to digest the insect and absorb its nutrients."

PITCHER PLANTS

VENUS FLYTRAPS

"Can we get some meat-eating plants for our yard?" Nick asked eagerly.

"I don't think I want any," Christie gulped. "They sound mean."

"They don't eat insects to be mean," Dad assured her. "God gave them the ability to capture and eat insects because that's the only way for them to get the nutrients they need."

"That's right," Mike said. "Pitcher plants grow in bogs where the soil has few nutrients. Most plants aren't able to grow there, but carnivorous plants can because they absorb nutrients from the insects they capture."

"Isn't the venus flytrap a carnivorous plant, too?" Dani asked. "It has leaves that snap shut when a fly lands in them."

"You're right, the venus flytrap is another carnivorous plant," Dad said. "It grows only in the southern United States, not in the boreal forest, but it captures insects for the same reason."

SUNDEW

"Because it grows in poor soil and needs another source of nutrients?" Dani asked.

"Exactly," Dad confirmed.

Fish that hatch in fresh water, live in salt water, and return to lay their eggs in fresh water are called **anadromous** (uh-NA-druh-muhs). This term comes from the Greek words *ana + dromos*, which mean "upward running," and refers to the fact that they travel upstream to lay their eggs.

## 9.8 Salmon

"Do I smell salmon?" Dad asked, making his way to the kitchen to wash his hands before dinner.

"Creamed salmon on toast!" Dani told him, arranging the toast on a serving dish. "It was Mom's idea—she said you want to tell us about the life cycle of the salmon, so we picked a special dinner for tonight."

"And I helped heat up the peas and butter the toast!" Christie said. "Dani wouldn't let me open the cans of salmon, though."

Dad led his family in saying grace, and soon everyone was enjoying a delicious Friday-night meal.

After he had eaten several bites and had complimented Dani and Christie on their cooking, Dad declared, "It's time to learn about salmon, the most interesting and delicious fish of all in my opinion. Most fish stay in either fresh water or salt water all their lives," he began. "But salmon are **anadromous**, which means they are born in freshwater rivers, spend most of their lives in the salty ocean, and then return to freshwater rivers to lay their eggs and die."

"Don't they get to see their babies?" Christie asked.

"No they don't," Dad replied. "God designed baby salmon, called fry, to be able to take care of themselves as soon as they hatch from their eggs. After spending some time in the freshwater rivers where they were hatched, the young salmon travel downstream to the ocean.

"Around this time, the salmon's liver changes so it can survive in salt water. Once it is used to living in the salty ocean water, it can travel for thousands of miles. The salmon eats smaller fish and other sea creatures, and spends from one to seven years growing in the ocean."

*Image Courtesy Green Water Fishing Adventures greenwaterguides.com*

50-POUND CHINOOK SALMON

"How big do salmon get?" Mike asked.

"There are several different kinds of salmon," Dad answered. "Some species don't grow much larger than three or four pounds, but Chinook salmon are often three feet long and weigh about 40 pounds. Chinook salmon are also called king salmon, and they hold the world record for the largest salmon ever caught. In 1949 a 126-pound Chinook salmon was captured in a fish trap in Alaska. It was over five feet long!"

"No way!" Nick exclaimed, almost choking on his mouthful of creamed salmon.

"If you think their size is amazing, wait until you hear about their strength and determination," Dad told him. "When the salmon are ready to lay their eggs, they return to the streams they were hatched in. Leaving the ocean, they swim up rivers and streams, fighting against the current. If their path is blocked by a waterfall, they jump up it, propelling themselves through the air with their powerful bodies. Some salmon have to travel hundreds of miles to reach their hatching grounds, and all this time they don't eat anything.

"Scientists aren't sure how the salmon find their way back to the same rivers they were hatched in. Most think they use their sense of smell. They have an extremely sharp sense of smell, much sharper than any dog, so they can remember what their home river smells like. Often, they even find their way back to the exact spot on the river where they hatched!

"When salmon reach their hatching grounds, they lay their eggs and cover them with gravel to keep them safe. Once they have laid their eggs, the salmon age quickly, and soon die.[2] Then in the spring, a new generation of salmon hatch."

Christie was still pondering the fact that the salmon don't eat anything during their journey upstream. "They don't eat anything at all while they're traveling?" she asked.

"Nothing at all," Dad said. "They are spending all their energy to reach their hatching grounds and lay their eggs before they die. Since they are in the last stage of their lives, they don't really need to eat."

"But I bet other animals eat them!" Mike commented. "I remember seeing pictures of grizzly bears catching salmon."

"Good point," Dad said. "When salmon travel upstream in the fall, they become food for bears, otters, eagles, and other boreal forest animals. Isn't God good to send His creatures such an abundant

---

[2] Some Atlantic salmon return to the ocean and survive for another year.

supply of food, right when they are trying to eat as much as they can to last them through the winter?"

"It's also good timing for the salmon," Mom added. "Salmon eggs take about three to five months to hatch. If the eggs are deposited in the fall, they will hatch in the spring, when the stream has the most food for young salmon. That way, more salmon will survive to adulthood and return to start a new generation of salmon."

"Salmon are definitely my favorite fish," Mike said. "May I please have a second helping of creamed salmon on toast?"

**Workbook pgs. 52-54**

# 10
## Chapter

# Forests: Temperate and Tropical

*I love to think of nature as an unlimited broadcasting station, through which God speaks to us every hour, if we only will tunc in.*

— George Washington Carver

## 10.1 Temperate and Tropical Forests

"Dad, can we study a warmer biome this week?" Dani asked. "I have 'virtual' frostbite and pneumonia after our road trips through the tundra and the boreal forest!"

"Yes," Christie agreed. "And I'm getting tired of drawing snow boots and coats."

"I was planning on taking you all on a virtual trip to the rainforest," Dad answered. "The temperature never drops below 60 or 70 degrees there. Will that cure your virtual frostbite?"

"That sounds much better," Dani assured him. "My toes are thawing out just at the thought of tropical weather."

"Is the rainforest where boa constrictors live?" Nick asked with excitement.

ANACONDA

"Yes, and anacondas," Mike said. "The anaconda is the biggest snake there is."

Christie huddled closer to Dani and announced that she was going to bring "snake spray."

"I'm not sure if there is such a thing," Dani told her, "but I won't let any snakes get near you."

"It's true there are snakes in the rainforest," Mom said, "but they aren't the main creatures there. South America is known as the 'bird continent' because there is such a variety of birds in its rainforests."

"That's right," Dad said. "And it's the South American rainforest that we'll be visiting. The Amazon rainforest in South America is the largest in the world. So what do you think we should pack?"

"Umbrellas," Christie said. "It's a *rain*forest."

"A machete!" Mike exclaimed.

"What's a machete?" Nick asked him.

"A machete is a kind of hatchet or knife that explorers use to hack their way through the jungle," Mike explained.

"Climbing gear would be useful," Mom suggested.

"Climbing gear?" Dani asked. "Are there mountains in the rainforest?"

"Yes, rainforests can grow on the sides of mountains," Mom said. "For instance, some of the rainforests in South America are located high in the Andes Mountains, and are blanketed in low-lying clouds and mists. I was thinking of using the climbing gear to climb trees, though. Most of the activity in a rainforest occurs in the tree tops."

"That sounds like fun!" Mike grinned. "Nick and I will definitely pack climbing gear in our suitcases!"

"I'm going to bring binoculars, too," Dad said. "There's nothing more useful than binoculars when it comes to birdwatching!"

"Should we bring bug spray again?" Dani asked. "Don't some of the mosquitoes in the rainforest carry malaria?"

"You're right, they do," Dad said.

Crown Victoria Pigeon

Emerald Toucanet

Black-throated Trogon

Bird of Paradise

"Why don't we bring mosquito netting?" Mike suggested.

"Wonderful idea!" Dad said. "I wish we'd thought of doing that in the tundra. Bringing mosquito netting on our tundra road trip would have saved me a lot of itching!" he exclaimed, scratching an imaginary bite. "Some scientists estimate that there are actually more mosquitoes in the tundra than in the rainforest."

GIANT ANTEATER

"So the rainforest doesn't have as many bugs as the tundra?" Christie asked hopefully.

"The rainforest has just as many, and more. I said the rainforest doesn't have as many *mosquitoes* as the tundra does," Dad explained. "It has thousands of other types of insects. One researcher identified 200 different species of ants on a single tree!"

"There are so many insects in the rainforest that some mammals eat nothing but insects," Mom added. "The giant anteater is a good example. This mammal tears open ant nests with its claws, then pokes in its long nose and uses its two-foot-long tongue to lick up the ants."

"That's gross!" Nick exclaimed.

"A lot of people don't think so," Mom informed him. "Insects are a delicacy in many cultures. I've seen advertisements for 'slow-roasted giant water bugs' or 'crunchy crickets and tarantulas in sweet-and-sour sauce.' Don't you think we should try some for dinner this week?"

## 10.2 Temperate Forests

"Is everyone's suitcase packed?" Dad asked. "Then let's take a look at the map and trace our progress so far.

"We started in the freezing arctic tundra, where the permafrost and the short growing season makes it impossible for trees to grow. The arctic tundra is near the top of the globe, so it experiences 24-hour days in the summer and 24-hour nights in the winter.

"Then we travelled south through the Canadian boreal forest. Trees are able to grow in this biome, because the ground thaws in the summer instead of remaining frozen as permafrost. The summers are still very short, though, and the long, cold winter is a major challenge for both plants and animals. Most of the trees are evergreen, so they can begin photosynthesizing as soon as the snow melts, instead of having to grow new leaves first.

"When we travel farther south, we come to temperate forests. These are the forests near our home. Before it was settled, all of Europe was a temperate forest, and temperate forests still cover much of the United States."

"A lot of our favorite stories and folk tales are set in temperate forests," Mom remarked. "'Hansel and Gretel,' for instance, and *Little House in the Big Woods*."

"Was Robin Hood's forest a temperate forest?" Mike asked.

"It sure was," Mom affirmed. "And so are the woods in *The Chronicles of Narnia* and in *The Lord of the Rings*."

HANSEL AND GRETEL

Raccoon

"In the boreal forest, the year is divided into a long winter and a few, short months of spring, summer, and fall," Dad continued. "But in the temperate forest, the year is divided into four equal seasons. Plants grow during the summer, and also during much of the spring and fall.

"A major difference between the boreal forest and most temperate forests," Dad explained, "is that there are more deciduous trees in the temperate forest. Deciduous trees lose their leaves in the winter. The growing season in the temperate forest is much longer than it is in the boreal forest, so the trees have plenty of time to grow new leaves in the spring.

White-tailed Deer

"The temperate forest also receives more precipitation than the boreal forest does," Dad remarked. "Temperate forests receive about 30-50 inches of rain and snow each year. Since the weather is warmer, fallen leaves and branches quickly decompose, so the soil of the temperate forest is rich and fertile."

Dad paused, and then asked, "Who can tell me what sort of animals live in the temperate forest? Remember, the woods near our house are part of a temperate forest."

"Raccoons!" Christie exclaimed.

"Squirrels and blue jays!" Nick shouted.

"White-tailed deer and red foxes," Mike said.

"Black bears and mountain lions in the wilder parts of the temperate forest," Dani suggested.

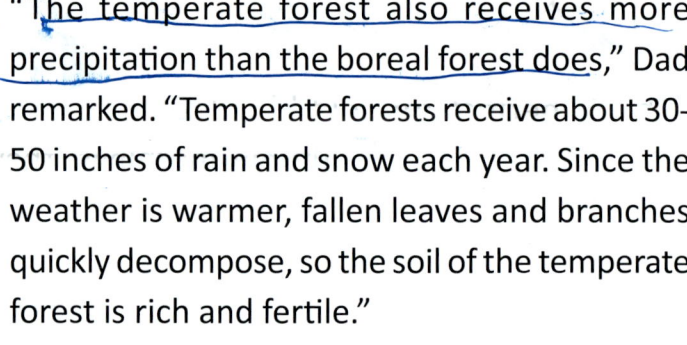

Spring Peeper

"That's a great list!" Dad said. "I'll add lizards, salamanders, frogs, and snakes. The temperate forest is the first biome we've studied where there is a significant number of reptiles and amphibians. These cold-blooded animals need the warmth of the Sun to remain active. In the winter, they have to hibernate underground to stay alive. The longer the winter, the harder it is for them to survive."

"I can't say I'm fond of snakes," Mom laughed, "but I find it amazing how God provides for them in the temperate forest. Have you noticed that in a deciduous forest the forest floor receives the most sunshine during the coldest seasons? In the summer, the forest floor is shaded by the trees' foliage, but in the autumn, the leaves fall from the trees, allowing more sunlight to reach the forest floor.

"This is perfect for cold-blooded critters!" Mom continued. "On chilly, autumn mornings, they can warm themselves on sunny rocks and stumps. Then, when winter arrives, they hibernate in burrows deep underground."

"And in the spring, the animals on the forest floor get a lot of sunshine because the trees are still growing new leaves!" Dani exclaimed.

"Exactly!" Mom said.

"Now that we've taken a quick 'drive' through the temperate forest, it's time to head on down to the Amazon rainforest!" Dad announced.

"Let's go!" Nick shouted, eager to learn about boa constrictors and anacondas.

The average summer temperature in the temperate forest is about 75°F. Winter temperatures range from -20° to 40°F.

Workbook: Temperate Forest Profile, pg. 55

In forests, most wildflowers bloom in the early spring. Do you think they might be taking advantage of the extra sunshine that is available before the trees' foliage is fully developed?

# Amazon Rainforest

THE BREVES NARROWS ON THE AMAZON RIVER, BRAZIL

SILVER AROWANA

CAIMAN

The Amazon rainforest is named after the Amazon River, which is the second longest river in the world, and the largest river when judged by volume. Each day, the Amazon River pours more than four trillion gallons of water into the Atlantic Ocean.

Every year during the wet season, the Amazon River floods its banks and spreads for miles into the surrounding forests. Some areas in the Amazon rainforest remain under 30-40 feet of water for eight to ten months!

In these flooded forests, tall trees stretch their branches above the water while fish and crocodile-like caimans swim among their trunks. Many of these fish are fruit eaters, and wait under the trees to feast on the fruit when it drops into the water. Another fish, the arowana, eats the insects on low-hanging branches. With perfect accuracy, the arowana will leap three feet out of the water to snatch a delectable bug!

©Emily S. Damstra

### CRESTED GREEN BASILISK

The crested green basilisk lizard is also known as the Jesus Christ lizard, because of its remarkable ability to run across the surface of a calm body of water when escaping from predators. It does this by spreading its long toes and racing over the surface at a speed of about five feet per second. As it slaps its feet onto the surface, tiny bubbles of air are caught beneath its feet, keeping it afloat until it takes its next step.

## 10.3  Tropical Rainforests

Tropical rainforests can be found in South and Central America, Africa, and Indo-Malaysia, a region that includes parts of China, India, the Philippines, Indonesia, and New Guinea. The Amazon rainforest in South America is the largest rainforest on Earth.

Rainforests are unique for the great diversity of plant and animal life that they contain. Tropical forests cover only 7% of the globe, but they contain about 50% of the plant, animal, and insect species in the world. For instance, the Mississippi River system is home to about 250 species of fish, but the Amazon River system provides habitats for approximately 2,000 different species of fish! Seven hundred tree species were identified in only 25 acres of rainforest in Borneo—the same number as the tree species in all of North America.

Tropical rainforests are situated near the Equator, so the temperature stays near 75°F all year round. The temperature hardly even varies between day and night. Unlike temperate forests, tropical rainforests do not experience four seasons. Instead, the year is divided into a wet season and a dry season. The rainfall is extremely heavy in the wet season, and relatively light in the dry season. Every year, the rainforest receives about 100 inches of precipitation, with some areas receiving as much as 400 inches. Not surprisingly, the air is extremely humid and the sky is often cloudy or filled with rain.

THE BUTTRESSED ROOTS OF THE CEIBA TREE

**Epiphytes** (EP-uh-fahyts) are plants that grow on trees or other plants, but do not get their nutrition from those plants.

ORCHID

EPIPHYTES

The abundant rainfall can be a challenge to the plants in the rainforest. The constant, heavy rainfall washes nutrients and minerals out of organic material. The mineral-filled rain then flows away into streams and rivers, or seeps deep into the ground, farther than the trees' roots can reach. Because of this, the soil in the rainforest is thin and poor, unlike the deep, rich soil in temperate forests. Rainforest plants survive by reabsorbing the nutrients in organic materials as quickly as possible, before the rain can wash the nutrients away. Trees in the rainforest often have extensive root systems close to the surface of the soil, which helps them absorb nutrients quickly.

In boreal and temperate forests, the growth of trees is limited by freezing winter temperatures and yearly seasons of drought. In the warm, humid rainforest, the temperature and amount of precipitation are not limitations, and thousands of different plant species grow and flourish, including the cacao (cocoa) tree. In fact, the rainforest climate encourages so much growth that rainforest plants compete with each other for sunlight, nutrients, and room to grow.

The main challenge for rainforest plants is getting enough sunlight. To compete with the trees around it and obtain as much sunlight as possible, each tree grows as many leaves and branches as it can. As a result, the foliage of rainforest trees is so dense that the forest floor is always dark and dim. In fact, it is difficult for small plants to grow on the forest floor because they don't receive enough light. Because of this lack of sunlight, the ground floor of a rainforest is relatively empty compared with the floor of a temperate forest. Most small plants and flowers grow high above the ground, on the branches of tall trees! Here they soak up the sunlight they need for photosynthesis, and absorb nutrients from the mosses, lichens, and fallen leaves that collect on the tree's branches. Plants that grow on trees or other plants are called **epiphytes**. Orchids are the most famous and beautiful epiphytes.

Most epiphytes are not harmful to the trees they grow on, but the strangler fig is a parasite that eventually kills its host tree. When the seed of a strangler fig is deposited on a tree, usually in bird droppings, it begins growing just like other epiphytes. Unlike most epiphytes, the strangler fig sends out dozens of long roots that grow along the tree trunk down to the ground. Through these roots, the strangler fig receives enough nutrients to grow large branches and leaves. Eventually, the strangler fig becomes so large that its spreading foliage deprives the host tree of sunlight, killing it. When the trunk of the original tree has rotted away, the strangler fig remains standing upright, supported by its lattice of strong roots. The competition between a strangler fig and its host tree demonstrates what a critical resource sunlight is in the rainforest.

Two species of strangler figs

FICUS VIRENS

FICUS WATKINSIANA

The **emergent layer** in a rainforest is a scattered layer of very tall trees.

The **canopy** is the densely-packed "roof" of the rainforest, the second layer from the top.

The **understory** in the rainforest is the dim, shady layer directly beneath the canopy.

## 10.4 Layers of Vegetation

There are four layers of vegetation in the rainforest. The top layer is called the **emergent layer**. Emergent trees are scattered throughout the rainforest, and raise their branches high above the trees around them. They usually stand 100-120 feet tall, and provide look-outs for harpy eagles, hawks, and other birds of prey.

The next layer is called the **canopy**. Viewed from above, the canopy looks like an endless, green carpet of leaves and branches. The trees in the canopy grow so close together that they block most of the light from the plants below them. Fueled by the intense supply of sunlight, canopy trees produce an abundance of leaves, fruit, and nuts, which provide food for countless creatures. The canopy is full of animals, including birds, bats, monkeys, silky anteaters, sloths, tree frogs, spiders, beetles, ants, butterflies, and other insects.

Below the canopy is the **understory**. The trees in the understory receive only a small amount of light, so they grow large, dark leaves to soak up what little sunshine penetrates through the canopy. The understory includes vines and small trees, as well as the saplings of canopy trees. This dim region of the rainforest is occupied by jaguars, leopards, snakes, lizards, squirrels, and many other creatures. Insects, of course, are found in every layer of the rainforest.

The fourth layer of the rainforest is the dimly-lit forest floor. Although there is relatively little vegetation on the forest floor, the fruit, nuts, and leaves that fall from the trees provide food for deer, peccaries (wild pigs), rats, agoutis, and ground-loving birds. The layer of decaying topsoil on the forest floor is home to millions of creeping insects, which in turn provide abundant food for the frogs, lizards, and snakes that hide in the fallen leaves and branches. At night, large cats such as the ocelot hunt for rabbits, birds, and other small animals.

OCELOT

## 10.5 Day and Night in the Rainforest

"Before we learn about the different ways rainforest animals live and work together," Mom said, "let's take a virtual walk through the rainforest.

"Imagine that Nick has landed his airplane in a clearing in the middle of the Amazon rainforest. We take off our seatbelts, get out of the plane, and walk to the edge of the clearing. The forest in front of us is dense and crowded. Mike has to get out his machete and chop away the branches and hanging vines before we can enter the forest."

Mike grinned and swung an invisible machete, but Dani looked

Workbook: Tropical Rainforest Profile, pg. 56

Research Assignment, pg. 56

 Workbook pg. 57

The skin of some tropical salamanders contains a poison that paralyzes the jaws of any snakes that attempt to eat them. The salamander can then walk out of the snake's mouth!

181

Black-capped Squirrel Monkeys

Blue-tailed Emerald Hummingbird

Leaf-cutter Ants

Malayan Tapir

puzzled. "I thought there weren't many plants growing on the forest floor," she said. "Why does Mike need a machete?"

"Because right now we are still near the clearing, where the forest floor gets plenty of sunlight," Mom explained. "Wherever there is a gap in the canopy—on the edge of rivers and clearings, for example, or when a tree falls—the forest floor is covered with lush undergrowth. But deeper in the forest, the canopy blocks most of the sunlight, so the undergrowth is much less dense. A few dozen yards into the rainforest, we are surrounded by hundreds of tree trunks reaching high into the air. The trunks of canopy trees stand like massive pillars holding up the roof of the forest.

"As we walk along, we start to notice the many different creatures around us. We hear hundreds of toucans and scarlet macaws shrieking and calling to each other as they fly from branch to branch in the canopy far above. Nick spots a family of squirrel monkeys swinging in the vines, and Christie sees a shimmering, blue-and-green hummingbird.

"Walking farther, we hear the snuffling and rooting sounds of a peccary, or wild pig, as it forages on the forest floor. Looking between the trees, we notice several agoutis burying brazil nuts in the ground. Dani sees a bright blue butterfly, and Mike discovers a trail of leaf-cutter ants carrying bits of leaf to their nest. The more time we spend in the rainforest, the more creatures we notice," Mom said.

"And when I persuade you to spend the night in the rainforest," Dad added, "we see even more animals. The rainforest is at least as active at night as it is in the

BLACK JAGUAR

GREEN TREE PYTHON

PINK DOLPHIN

daytime. When we shine our flashlight through the forest, we see dozens of glowing eyes, and fireflies fly among the trees flashing their lights.

"Bats flit through the forest and a boa constrictor slithers along a branch. Crouched on a large branch, a jaguar silently lies in wait for a grazing deer or Brazilian tapir. Pollinating moths fly from flower to flower. As dawn approaches, we hear a group of night monkeys, or dourocouli, noisily swinging through the trees in search of fruit. As they travel, the monkeys call loudly to each other so the group does not become scattered."

"It doesn't sound so impossible anymore for tropical forests to contain half of the world's species!" Mike exclaimed.

HARPY EAGLE

GREEN IGUANA

183

**Biodiversity** is the variety of plant and animal species in a biome.

**Specialization** describes the way different creatures obtain their food in different ways, thus reducing competition.

Brazil nuts grow inside a double shell. The hard, outer shell is about the size of an orange. About a dozen brazil nuts in secondary shells are arranged within the outer shell like the segments of an orange.

Zanzibar red colobuses

## 10.6 Specialization

The rainforest contains more **biodiversity** than any other biome. This means that there is a greater variety of creatures in the rainforest than anywhere else. Since millions of insects, reptiles, amphibians, birds, fish, and mammals live close together in the rainforest, every creature has a very specific niche and habitat. Some animals are active only during the day, and others are active only at night. One species of monkey might live only in the canopy while another makes its home in the understory.

Many animals "stake a claim" on a particular type of food as well as claiming a particular territory or time of day. For instance, very few mammals eat ants, so the giant anteater has little competition as it breaks into ant nests to lick up its favorite meal. Different species of anteaters avoid competing with each other by foraging for food in different places; the silky anteater is about the size of a house cat and lives solely in trees, while the giant anteater finds its food on the forest floor.

AGOUTI

SWORD-BILLED HUMMINGBIRD

WHITE-TIPPED SICKLEBILL

The agouti, a rodent, is one of the only animals with teeth sharp enough to bite through the hard, outer shells of Brazil nuts. Since other animals cannot eat the nuts before the agouti gets to them, the agouti has few competitors. The bill of a sword-billed hummingbird is as long as the bird itself, which allows it to reach the nectar deep within tube-like flowers. Similarly, the white-tipped sicklebill has a curved bill that fits perfectly into *Heliconia* flowers.

Because God equipped these creatures with special noses, tongues, teeth, and bills, they are able to find unique sources of food, which reduces competition with other animals. This arrangement, called **specialization**, makes it possible for many rainforest species to live and flourish in a small area.

## 10.7 Predator and Prey

Besides finding food for themselves, rainforest animals also try to avoid becoming dinner for other animals. Large cats such as jaguars and tigers are at the top of the food chain, along with giant snakes such as anacondas and pythons. The anaconda kills its prey by suffocating it in its powerful coils,

BENGAL TIGER

Green Anole

Brown-Throated Sloth

Leaf Mimic Katydid

Green Tree Python

and then swallows its victim whole. Some pythons are equipped with heat-sensitive pits on their noses that allow them to track down prey by detecting its body heat.

To avoid being eaten, many creatures are equipped with incredible camouflages. For instance, the anole is a lizard that can change its skin color to match its surroundings. The sloth moves so slowly that green algae begin to grow in its fur, which helps it blend in with its surroundings. The leaf mimic katydid looks exactly like a dead leaf.

Of course, rainforest predators are also camouflaged. The bright green body of the green tree python helps it hide from its prey, and the spotted fur coats of jaguars blend in with the dappled forest floor.

Some animals have colors that make them stand out instead of providing camouflage. Poison dart frogs contain a deadly poison, so they do not have to hide from predators. Instead, their bright colors warn other animals away. The hawk moth frightens its predators away by flashing the spots on its wings to make them look like the eyes of a large, dangerous bird.

Jaguar

Blue Poison Dart Frog

Hawk Moth

*Azteca alfari* ants care for their eggs inside the hollow stems of the *Cecropia* tree.

## 10.8 Interdependence in the Rainforest

"Our theme this evening is interdependence in the rainforest," Dad said. "Mom tells me you researched some amazing examples this afternoon. Who wants to go first and tell us what you learned?"

"I do!" Nick exclaimed. "Mike and I learned how trees and ants work together."

"Really?" Dad said. "Tell us about it."

"*Cecropia* trees grow hollow branches for ants to live in," Nick explained. "The ants like making their nests there because the tree feeds them. The tree likes having the ants living in it, because if an animal comes to eat the tree's leaves, the ants attack it!"

"How can ants protect a tree?" Dani asked.

"Would you like to be bitten by hundreds of stinging ants?" Mike asked his sister. "The ants protect their tree so fiercely that they even chew off vines and plants that try to grow on their tree."

"Nick mentioned that the tree feeds the ants," Dad said. "Do the ants eat the tree's leaves or something?"

187

BROMELIAD

"No," Mike said. "That would be just as bad as having another animal eat its leaves. At the base of its leaves, the *Cecropia* tree grows little packages of carbohydrates, called Müllerian bodies, for the ants."

Dad shook his head in amazement. "I have to say, ants and trees working together is the most impressive example of facilitation I've ever heard of."

"Christie and Dani learned about another example of facilitation," Mom remarked.

"Great!" Dad said. "Tell us about it."

"We learned about tank bromeliads," Dani said. "Bromeliads are plants in the pineapple family, and a lot of them are epiphytes."

"That means they live and grow right in the tree tops," Christie explained to her siblings.

CROWN TREE FROG

"Tank bromeliads grow thick leaves in tight clusters," Dani continued. "The leaves in each cluster overlap at the bottom, forming a water-tight tank. When it rains, the bromeliad's tank fills up with water. Some bromeliads can hold as much as two gallons of water!"

"Then when the plant is thirsty, it just takes a drink!" Christie said.

"Why do they need to store water in the rainforest?" Mike asked. "Don't they get more water than they need?"

"Actually, it can get quite dry up in the tree tops," Dani

said. "But the water-tanks of bromeliads don't just collect water—they also collect organic material like dead leaves and insects."

"Oh, I get it!" Mike exclaimed. "The organic material in its tank is where the bromeliad gets the nutrients it needs."

"Right!" Christie said.

"Yes, but it's more complicated than that," Dani said. "Organic material contains lots of nutrients, but bromeliads can't use them until they have been decomposed. This is where facilitation comes in. Tank bromeliads are a favorite habitat for algae, bacteria, mosquito larvae, tadpoles, salamanders, and other small animals," Dani explained, glancing at her notes.

"All of these creatures," she continued, "especially the algae and bacteria, help to decompose the organic material into a form the bromeliad can use."

"And in exchange, the brow-mee-leeadd gives them a nice place to live!" Christie exclaimed. "See the picture I drew?" She held out a well-crayoned piece of drawing paper covered with neon-green leaves, aquamarine water, and sunset-orange tadpoles.

Once Christie's picture had been properly admired, Dani summarized, "So tank bromeliads provide little critters with a place to live, swim around in, and lay their eggs, and the little critters provide the bromeliad with food."

"I'd like to live in the rainforest," Nick declared, impressed by the fact that bromeliads are part of the pineapple family. "I could spend all my time eating pineapple and playing with snakes!"

 Experiment #25

 Workbook pgs. 58-60

# Chapter 11

# Grasslands and Deserts

Nature is the art of God.
— Dante Alighieri

## 11.1 Grasslands and Desert

"So far we've visited the tundra biome and three different forest biomes," Dad said. "You might not have noticed, but we've been studying the biomes in a particular order.

"We began with the tundra, which has very little precipitation and very cold temperatures," Dad explained. "Then we went to the boreal forest, which receives more precipitation than the tundra and has slightly warmer temperatures. Next, we visited the temperate forest, which is even warmer, and receives more precipitation. Finally, we went to the rainforest, which is really hot and really wet. Do you see the pattern here?"

"We started with dry and cold and worked our way to wet and hot," Mike said.

"Exactly. Now we're going to leave the hot and wet rainforest and travel to biomes that are hot and *dry*. We're going to take a trip to the grasslands and the desert."

"Oh good!" Nick said. "I can wear shorts!"

"And sunscreen," Mom commented. "I have two suitcase drawings on the kitchen table for you and Christie," she added. "Why don't we go pack our 'suitcases' for the grasslands and desert, while Dad introduces Mike and Dani to the weather and soil of these biomes?"

"I'll beat you there!" Nick shouted.

## 11.2 Prairie

"There are two types of grassland," Dad explained to Dani and Mike, "temperate grasslands and tropical grasslands. The prairies that we see when we drive through the Midwest to visit your cousins, Josh and Hanna, are temperate grasslands. Just like temperate forests, temperate grasslands experience four seasons with a hot summer and a cold winter. Temperatures in temperate grasslands can be as high as 100°F in the summer and as low as -40°F in the winter!

"Like temperate forests, temperate grasslands have very fertile soil. But instead of being covered with trees, temperate grasslands are covered with beautiful grasses and wildflowers."

"If the soil is just as fertile, why don't trees grow there?" Dani asked.

"The main reason there are few trees in the prairie is fire," Dad said. "Prairie fires used to be quite frequent in the Midwest, even more frequent than in the boreal forest. Seedlings can't grow into mature trees if they are burned by prairie fires every few years."

"Don't fires kill grass, too?" Mike asked.

"Surprisingly, they don't," Dad replied. "Most plants grow from the tips of their branches, but grass grows from its base. Prairie fires destroy the blades of grass, but they don't hurt the stems and

The North American prairies, the South American pampas, and the Russian steppes are all temperate grasslands.

BLACK-EYED SUSAN

BIG BLUESTEM PRAIRIE GRASS

The grazing of large animals is another reason trees are scarce in the prairie. Like prairie fires, the nibbling of hungry antelopes and deer kills young trees, but does not harm the grass, which soon sprouts again from its deep roots.

roots, because these are protected underground. Prairie grasses grow very deep roots—some up to 20 feet deep! As soon as the fire has passed, these roots send up fresh, new blades of grass. A few weeks after a fire, the prairie is once more green and full of life."

"Hey! That's the way crab grass and quack grass act in the garden!" Mike said. "If I just pull off the top of the plant, it grows back stronger than ever. I have to pull out the roots as well as the leaves, or the weed will just grow back again."

"I'm impressed that you made that connection," Dad said. "You are right—grass grows back after a prairie fire for the same reason that grass-like weeds grow back if you leave their roots in the ground."

"I have another question," Dani said. "You said that the prairie has very fertile soil. Can you tell us why?"

"The richness of prairie soil is mainly due to rainfall and temperature," Dad answered. "Do you remember how in the

BISON (BUFFALO)

PRAIRIE DOG

PRONGHORN ANTELOPE

The prairies were once home to thousands of bison (buffalo), deer, and pronghorn antelope, which grazed on the abundant grass. Much of the original grassland has been developed into productive farmland, but herds of deer and pronghorn antelope still roam the prairie. Small animals, such as prairie dogs, rabbits, ground squirrels, and snakes, are also abundant in the temperate grasslands of North America.

rainforest, the abundant rain carries the nutrients deep into the ground, farther than plants can reach? This doesn't happen in temperate grasslands. Prairies receive about 20-30 inches of precipitation per year, which is just enough rain to moisten the soil so dead organic material can decompose easily, but not so much that the nutrients are carried out of reach of the plants.

"Decomposition is also aided by the warm temperatures during much of the year," Dad said. "In cold climates, dead organic materials decompose only slowly, so plants are unable to use the nutrients in these materials. The long, warm, prairie summers promote decomposition and ensure that the nutrients in organic materials are released back into the soil. Since the prairie's soil is so fertile, much of it has been converted into farmland for growing wheat, corn, and other crops."

"Like Josh and Hanna's farm," Mike commented.

"Exactly. The prairies in the midwestern United States are known as the 'breadbasket of the nation.'"

## 11.3  Savanna

"Grasslands can also be found in tropical climates," Dad continued, "where they are called tropical grasslands or savannas. Like tropical rainforests, tropical grasslands remain relatively warm all year round. Average temperatures range from 60°F to 95°F, although some areas experience more extreme temperatures. Tropical grasslands consist of scattered trees surrounded by fields of grass.

"Tropical grasslands receive more precipitation than temperate grasslands—about 20-60 inches per year—but this rainfall is

Although they may not look like the grass in your front yard, wheat and corn are domesticated species of grasses. No wonder they grow so well in the temperate grassland!

**Workbook: Temperate Grassland Profile, pg. 61**

South American savannas receive up to 100 inches of rain per year, and undergo major flooding during the wet season.

Australian savannas are home to many marsupials, including kangaroos and opossums. **Marsupials** (mahr-SOO-pee-uhlz) are mammals that have pouches for raising their young.

KANGAROOS

BLACK-MANED LION EATING A KUDU

highly seasonal. During the wet season, which occurs in the summer, rain falls in torrents. But during the dry season, tropical grasslands receive very little precipitation, if any. Surviving the dry season is a major challenge for both plants and animals.

"There are tropical grasslands in South America and Australia, but we are going to 'visit' the African savanna. The African savanna is home to millions of grazing animals. Everyone knows about zebras, giraffes, and Thomson's gazelles, but wildebeests, gerenuks, kudus, and impalas are also grazers in the African savanna. Lions, leopards, cheetahs, and African wild dogs prey upon these large grazers, while elephants, rhinoceroses, and ostriches browse for food in the trees, bushes, and grass."

"Nick will enjoy meeting giraffes and elephants on our 'virtual' road trip," Mike grinned.

"He certainly will," Dad agreed. "The biggest, most impressive animals are not always the most important, though. There are two creatures in the savanna that most people would not consider interesting at all, and probably would not want to meet. But these creatures have such important niches that if they didn't do their job, life in the savanna might not be possible. Can you guess what they are?"

"Probably some sort of insect," Dani joked, rolling her eyes. "None of them seem really interesting to me, but they're all over the place in just about every biome we've visited."

GIRAFFES

"You're right!" Dad said. "The creatures I'm thinking about are both insects. One is the termite and the other is the dung beetle."

"Umm, I can see why most people wouldn't want to meet them," Dani said.

"Wait until you learn more about them," Dad told her. "Dung beetles get their name because that's what they eat: dung. And they don't just eat dung; many of them collect it. The dung beetle shapes a large piece of dung into a ball and rolls it along the ground until it finds a good spot for digging. Then the beetle buries the dung, and either eats it later on, or lays its eggs next to it. The buried dung will serve as food for the beetle larvae."

Dad paused, looking at the expressions on Dani's and Mike's faces. "I know it sound disgusting, but the dung beetle is

A gerenuk stands on its hind legs to reach the leaves on the top of the bush.

IMPALAS ON THE RUN

197

DUNG BEETLE

The secretary bird spends most of its time on the ground, striding through the grass in hunt of prey, which ranges from grasshoppers and lizards to hares and venomous snakes. The three-foot-tall secretary bird overpowers small mammals and snakes by stamping and jumping on its prey until it is stunned or dead.

extremely important to the savanna biome. Dung is a powerful fertilizer, and the industrious dung beetle distributes it all over the savanna. This is especially necessary during the dry season. When there is rain, dung tends to dissolve and soak into the soil by itself. But in dry weather, dung simply hardens and dries, often killing the grass beneath it.

"Or rather, that is what would happen if it weren't for the dung beetle. Thanks to a lowly insect with a disgusting job, the nutrients found in animal dung are quickly buried in the ground, where grass and trees can absorb them to produce food for more animals."

"Okay, I guess dung beetles are pretty important," Mike admitted. "What about termites?"

"Like dung beetles, termites are 'soil enrichers,'" Dad explained. "Termites build huge, underground nests all over the savanna. Their tunneling mixes and loosens the ground, and their droppings enrich the soil.

"Many termite nests are topped by huge mounds, which are important structures in the savanna," Dad continued. "Eagles, cheetahs, and mongooses use them as look-outs, and smaller animals rest in their shade. A termite mound is a source of food for termite-eating aardvarks and birds such as the red and yellow barbet, and it is a back scratcher for elephants. Deserted termite mounds provide habitats for snakes, bat-eared foxes, warthogs, and other species.

"The savanna is dotted with thousands of these mounds, which can be more than fifteen feet high. The termites live in an underground nest beneath the mound, and the mound itself serves as an air-conditioning unit to keep their nest cool."

"No way!" Dani exclaimed incredulously. "An air-conditioning unit?"

"Yep. Termites are the most skilled architects on Earth. Even human architects can learn a lot from them. A termite mound is full of tunnels and holes that ventilate the underground nest. The mound acts like a chimney to carry away warm air and keep the nest at an even temperature.

"The ventilation systems in termite mounds are so sophisticated that some architects have begun to imitate termites when they design buildings. Recently, in the African country of Zimbabwe, a large shopping mall and office building was built without normal air-conditioning and heating systems. Instead, the inside

A family of mongooses sitting on a termite mound. The mongoose is famous for its skill in killing snakes, even the deadly cobra.

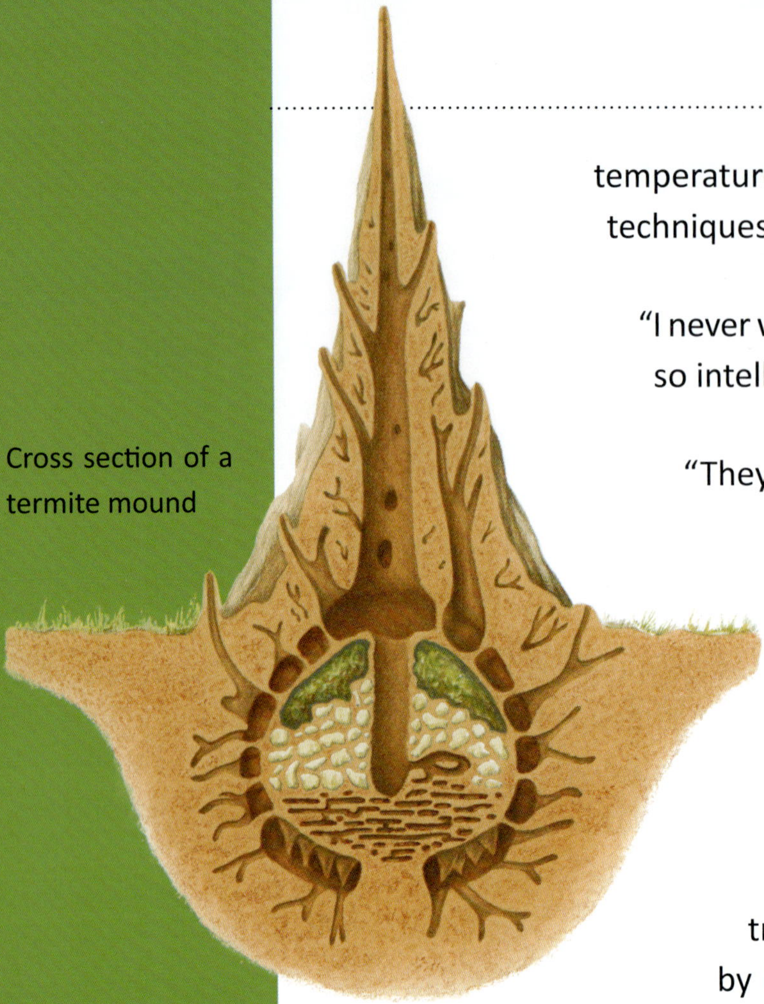

Cross section of a termite mound

Workbook:
Tropical Grassland
Profile, pg. 61

Workbook
pg. 62

temperature is controlled by the ventilation techniques found in termite nests."

"I never would have guessed termites were so intelligent!" Dani said.

"They aren't," Dad told her. "Termites are tiny, blind insects with hardly any brain at all. The million or so termites that build a termite mound aren't directed by their own intelligence, but by the instinct God gave them when He created them. Just as salmon travel upstream and birds fly south by instinct, so termites instinctively know how to build termite mounds. When we realize that their building plans were given to them by the Master Builder of all Creation, it's not really surprising that termites are able to build such sophisticated structures."

## 11.4  Desert

"So what did you pack in your suitcases for the desert?" Dad asked Nick and Christie the next evening.

"Sunscreen, jeans, sneakers, and sunglasses," Nick listed.

"I drew a big hat, a water bottle, and a coat," Christie said.

"A coat? I don't think you'll need one of those in the desert," Mike said.

"Yes, we do," Christie said. "It gets chilly in the desert at night, right Mommy?"

"Yes, the temperature at night can be 50 degrees lower than the temperature in the day," Mom said.

"That sounds impossible!" Mike said.

"Do you remember learning that the temperature doesn't change very much in the rainforest?" Mom asked. "This is partly because the heat from the Sun is trapped by the thick, humid air. Instead of escaping into outer space at night, heat rays from the warm ground are absorbed by the humid atmosphere.

"Unlike the air in the rainforest, desert air is very dry," Mom explained. "During the day, the Sun's rays beat down on the desert ground. With no water vapor in the air to block the Sun's rays, the desert becomes extremely hot in only a few hours. But once the Sun sets, the desert radiates much of the heat back into outer space and can become quite cool."

"This is why many desert animals sleep in the day and find food and water at night," Dad commented. "They are taking advantage of the cooler temperatures. Other desert animals are active in

Daytime temperatures in the desert are often 100-115°F. At night, the temperature can fall to 50-60°F.

ARCTIC HARE

BLACK-TAILED JACK RABBIT

ARCTIC FOX

FENNEC FOX

the day, but only in the early morning and evening.

"Desert animals have other ways of surviving the intense heat, too," Dad added. "Look at these pictures of the arctic hare and the black-tailed jack rabbit. Do you see how much thinner the jack rabbit's fur is, and how long its ears and legs are compared to the short ears and legs of the arctic hare? These are ways that God has given the jack rabbit to help it stay cooler."

"How can long ears help you cool off?" Nick asked skeptically.

"A rabbit's ears are full of blood vessels that carry its warm blood near the surface of its body," Dad explained. "The air that flows across the bare skin of the ear cools the rabbit's blood. The larger its ears are, the more blood it can cool at a time.

"By increasing or decreasing the flow of blood to their ears, jack rabbits can even control the amount of heat they lose," Dad continued. "In a similar way, the human body can reduce the flow of blood to the hands and feet in order to reduce heat loss. This is why our hands and feet often feel cold before any other part of our bodies."

"Also, slender bodies with long, narrow limbs naturally lose heat more quickly than chubby, round bodies," Mom added. "This explains why God designed arctic animals to have short legs and ears, while desert animals are often skinny, with long legs, tails, or necks."

"Like the long legs and necks of camels and ostriches!" Mike said.

"Exactly," Mom agreed.

## 11.5  Day and Night

"Imagine we are visiting the desert in the middle of summer," Dad said. "We just had lunch, and we decide to take a walk through the desert. Do you think we will see many animals?"

"No, the afternoon is the hottest time of the day," Mike said. "The animals will probably all be hiding in the shade."

"Yeah, or they'd get roasted," Nick commented.

"That's right. If we took a walk in the middle of the afternoon, we might see a jackrabbit standing in the shade of a cactus, a greater roadrunner with a lizard in his mouth, or a golden eagle gliding on the hot air rising from the desert ground, but other than that, the desert would probably look completely empty. Most desert animals rest in the shade or underground in the afternoon. Even snakes and lizards often bury themselves in the sand to escape the intense heat.

"So instead of taking our virtual walk in the afternoon, let's take one in the evening after dinner. Walking between the tall, saguaro cacti and bushy palo verde trees, we gaze in awe at a gorgeous sunset. As the desert becomes cooler, dozens of Mexican free-tailed bats begin flitting across the sky, and an elf owl emerges from its hole inside a saguaro cactus. In the distance, we hear the howls and yaps of coyotes.

"As we walk farther, Mike spots a night snake disappearing beneath a bush. Even though they are venomous, night snakes are not dangerous to humans because their venom is 'mild.' Still, the encounter reminds us to watch our step as we walk across the dry rocks and bare dirt of the desert floor. We don't want to step on a western diamondback rattlesnake!

"Dani suggests we sit quietly for a while to see if more animals

ROADRUNNER WITH LIZARD IN ITS MOUTH

Animals that are active at night are said to be **nocturnal**.

The five-inch elf owl is the world's smallest owl.

Barn owl

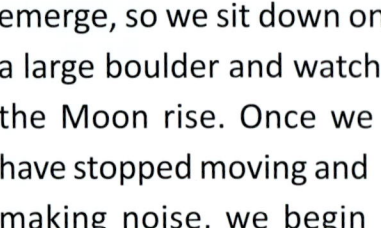
Kangaroo rat

emerge, so we sit down on a large boulder and watch the Moon rise. Once we have stopped moving and making noise, we begin to notice even more wild creatures. We hear a small rodent nibbling on seeds in the bushes behind us, but before we can figure out what it is, a barn owl swoops down on it silently. The animal hears the owl with its super-sensitive hearing and leaps out of reach, travelling nine feet in a single bound. Now we know the animal was a kangaroo rat, packing seeds into its cheek pouches. The owl flies away on silent wings to pursue some other prey.

"As the night grows colder, we begin to appreciate the warmth of the boulder on which we are sitting. A mule deer comes into sight, followed by two more. They graze on leaves and grass, picking their way delicately through the rough desert terrain. Suddenly, they raise their heads and freeze in alarm, looking towards the boulder where we are sitting. Nick sneezes and the deer dash away in long, graceful bounds."

Mule deer

"I did not sneeze!" Nick objected indignantly.

"Yes, you did," Mike said solemnly. "I heard you." Nick was ready to protest again, but Dad interrupted him.

"The full Moon has risen higher now, and we start to notice movement on the desert floor. Nick takes out his flashlight and shines it over the ground. We spot two or three lizards, several large beetles, and a Mexican red leg tarantula. The six-inch-wide tarantula pounces on one of the beetles and subdues it with a venomous bite."

Christie shrieked and pulled her feet up underneath her. "Don't worry," Dad assured her. "Most tarantulas are only venomous to insects and small animals, not to humans. In fact, a lot of people keep them as pets.

"Eventually we get sleepy, and take a long nap. We wake up as the Sun is rising, just in time to see a kit fox trot by with a mouse in its jaws. Dozens of birds are chirping and singing as they nibble on seeds and fruit or catch flying insects in the air. It doesn't take long for the desert to heat up, though, and soon most of the animals go into hiding again to wait for the cooler, evening temperatures."

"Then we get back on the road and turn on the air-conditioning, right?" Mike asked with a grin.

Mexican red leg tarantula

Workbook: Desert Profile, pg. 63

## 11.6  Did You Say Hot and Dry?

Strictly speaking, a desert is defined as any place that receives less than 15-20 inches of precipitation per year. By this definition, the tundra is a type of desert, since its average rainfall is 5-20 inches. For most people, however, the word "desert" calls up images of dry, cracked ground, a burning Sun, and parched plants and animals. In this section, we will study the hot deserts that

can be found in the temperate and tropical regions of the Earth.

Precipitation in hot deserts varies from 0-20 inches, so drought is a major challenge for both plants and animals. This challenge is increased by the unreliability of precipitation in the desert. A desert might receive double its normal rainfall one year, but no water at all for the next three years. So even in deserts with an *average* yearly rainfall of 15-20 inches, the inhabitants must be able to survive long periods of extreme drought.

Of course, our Creator, who created animals that can survive in the freezing temperatures and 24-hour nights of the tundra, is also able to equip animals and plants to live in the burning heat and drought of the desert. Desert animals are so well adapted to the scarcity of water in their desert home that some do not have to drink at all. Instead, these animals extract all the water they need from their food. Even animals that do have to drink regularly, such as kangaroo rats and large mammals, are able to survive with much less water than most animals.

Some insects, snails, reptiles, and amphibians escape the summer heat by going dormant. Just as animals in cold climates enter a dormant state—called hibernation—during the winter, so desert animals enter a dormant state—called **estivation**—during the summer. When animals hibernate or estivate, their body processes slow down, including their heart rate. In this dormant state, animals use much less energy, so they need hardly any

The highest temperature on the face of the Earth was 136°F, recorded in the shade in Africa's Sahara Desert. The desert ground can get much hotter, even reaching 180°F.

**Estivation** is a period of summer dormancy.

206

food or water. Animals usually estivate underground in burrows or holes.

Darkling beetle

Reptiles and insects have a special advantage in the desert because they do not sweat, and thus lose less moisture than mammals. For darkling beetles and Peringuey's adders in the coastal Namib deserts of Africa, being cold-blooded is an advantage, too. After their bodies have cooled during the night, the warm fog that moves inland each morning condenses on their bodies in tiny droplets. The adder then licks the moisture off his body, and the beetle does a "hand stand," lifting its rear end high into the air so the moisture rolls down its body towards its mouth.

Peringuey's adder

Birds are also well-adapted to desert conditions. Except for flightless birds like the ostrich, most birds can use their wings to fly to water sources and to escape the hot temperatures near the desert floor. The African sandgrouse will fly as far as 20 miles every morning to bring water to its young. When it arrives at a water source, the bird soaks its specially adapted breast feathers in the water and returns to the nest. There, the chicks eagerly suck the water from the adult's sponge-like feathers.

African sandgrouse

The desert air might be hot, but the surface of the ground is even hotter, sometimes reaching 180°F. Many desert animals, including camels, ostriches, jack rabbits, and even some kinds of lizards and ants, have extra-long legs that help them stay a little farther from the hot ground. The shovel-snouted lizard props itself up with its tail and performs a "thermal dance," standing on two feet as long as they can bear the heat, and then switching to its other two feet. Like

Shovel-snouted lizard

Peringuey's sidewinding adder

Saguaro cactus

many other snakes and lizards, the shovel-snouted lizard can also "swim" beneath the sand to avoid the hot surface of the ground or to escape from predators. Sidewinders thrust themselves across the sand in S-shaped loops instead of slithering like other snakes. This motion keeps them from slipping in the sand, and also helps them stay cool, since part of their body is always in the air instead of touching the burning sand.

## 11.7 Desert Plants

Just like desert animals, desert plants need special equipment to survive. In some ways, it is even harder for plants to survive than it is for animals. Animals can move out of the sun during the day, and can travel to find water. Plants must have other ways of surviving the heat and drought of the desert.

Some of the most well-known desert plants are the many varieties of cacti. Whenever a little bit of rain falls, cacti immediately soak up the moisture through their shallow, spreading roots. Then the cacti store the water in their thick, green stems. Of course, thirsty animals would love to eat a cactus for the water it contains. To protect cacti from jack rabbits, deer, and other herbivores, God gave cacti

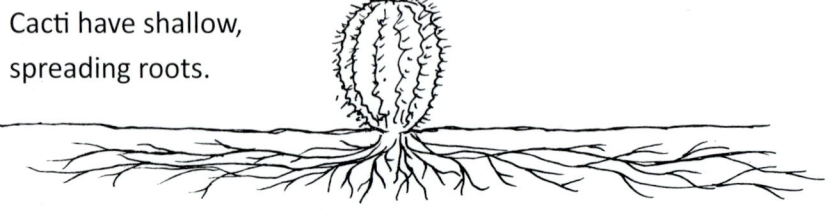

Cacti have shallow, spreading roots.

sharp spines. Cactus spines also reflect sunlight away from the plant, allowing the cactus to stay a little bit cooler.

Shrubs and trees cannot store as much water in their stems as cacti can, so they sink deep tap roots as far as 100 feet to reach underground water sources. Many of them are deciduous. In temperate forests, deciduous trees lose their leaves in the autumn. In the desert, deciduous trees lose their leaves in the summer. They do this to save water, since the part of a plant that loses the most water is the leaves.

Some plants, called **ephemerals**, don't even try to survive during the summer. During the wetter, cooler winter, these fast-growing plants sprout, bloom, and distribute their seeds. By the time summer arrives, ephemerals have already completed their life cycle. The mature plants wither and die, but their seeds survive through the summer to sprout and produce new plants in the fall.

All plants use sunlight to produce energy through photosynthesis. To carry out this process, plants must take in carbon dioxide through tiny holes in their leaves, called **stomata**. Unfortunately, when a plant opens its stomata to take in carbon dioxide, water is also able to escape. To help desert plants survive, God gave many of them the ability to store carbon dioxide. This allows them to take in carbon dioxide at night, when water evaporates less

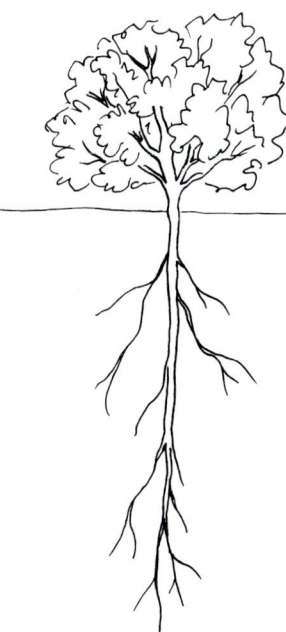

Desert shrubs grow deep tap roots to reach underground sources of water.

**Ephemerals** are fast-growing plants that complete their entire life cycle within a few weeks or months.

**Stomata** are tiny holes in a plant's leaves.

quickly. When morning comes, the plants close their stomata and complete the process of photosynthesis without losing water to evaporation.

Unlike rainforest plants, desert plants do not have to compete for sunlight—there is more than enough to go around! Instead, desert plants have to compete for water. If two desert plants grow too close together, there will not be enough water for both of them, and one will eventually die. This is why, in extremely dry deserts, large plants are always spaced a few yards apart from each other. Some plants, such as the brittle bush, even produce chemicals that are poisonous to other plants. This prevents other plants from growing near them.

Due to the scarcity of water, it is crucial that seedlings sprout at a distance from their parent plants. If a plant's seeds sprout right next to the parent plant, they will not be able to survive because the roots of the mature plant will absorb all the water. Some plants disperse their seeds by producing fruit. When animals eat the fruit, the seeds usually pass through the digestive tract unharmed and land in a new location. Then the seeds can sprout without having to compete with the parent plant.

Tumbleweeds have a unique method of dispersing their seeds. Mature tumbleweeds end their life cycle by becoming detached from their roots. The round, ball-like plant is then blown all about the desert by the wind. The rolling tumbleweed scatters its seeds wherever it goes.

The *Lithops* plant survives the desert climate by storing water in its stem, just as cacti do. Unlike cacti, *Lithops* do not have spines to protect them from thirsty animals. Instead, God gave these plants an amazing camouflage: small, knobby, and multicolored, a *Lithops* looks just like a large pebble. No wonder *Lithops* are known as "living stones!"

## 11.8 Conclusion

"Now that we've visited all seven biomes, can you tell me which is your favorite?" Dad asked.

"The rainforest!" Nick shouted.

"The tundra was my favorite place to visit," Dani said.

"I like our home forest best," Christie piped up.

"So do I," Mom agreed. "I love how the leaves turn color in the temperate forest."

General Sherman, a 2,500-year-old giant sequoia, is the largest living tree on Earth.

"The boreal forest was my favorite biome," Mike said. "But now that we've finished studying the biomes, I have a question for you, Dad."

"Ask away," Dad said.

"Which biome do Grandma and Grandpa live in? The mountains where they live aren't very far from us, but their climate seems a lot different."

"Good point!" Dad said. "Not every place on Earth will exactly match one of the biomes we have studied. For instance, there is a rainforest in California called a temperate rainforest, where giant sequoia trees grow. Some temperate forests are full of evergreen trees instead of deciduous trees, and some hot deserts actually receive snow in the winter.

ALPINE TUNDRA

BOREAL FOREST

TEMPERATE FOREST

"Mountains are an excellent example of why it is so difficult to categorize climates. The mountains where Grandma and Grandpa live are actually three biomes stacked on top of each other. Basically, when the **elevation** changes, the climate does too. Down in the foothills, the vegetation is similar to the deciduous trees in our own forest. At this elevation, the mountain's climate is that of a temperate forest. When we drive farther up the mountain, the air becomes cooler and the deciduous trees are replaced by evergreen conifers. At this elevation, the climate is similar to that

**Elevation** is the distance above sea level.

in the boreal forest, and the yearly snowfall is at least three or four feet. This is where Grandma and Grandpa have their cabin. Finally, if we drove even farther up the mountain, we would eventually reach the tree line. Above this point, it is too cold for trees to grow. This is called alpine tundra."

"You mean we could visit the tundra without going to the Arctic?" Dani asked.

"Well, alpine tundra isn't exactly the same as arctic tundra, but it is very similar."

"I think you'd appreciate the differences," Mom smiled. "There aren't swarms of insects in alpine tundra, because the soil is well-drained. It doesn't remain dark all winter, either. And since alpine tundra is found on top of mountains, there are often spectacular views."

"Let's go!" Dani exclaimed.

"I want to see polar bears!" Nick said.

"Do you think we can take a trip up there this summer?" Mike asked hopefully.

"There aren't any polar bears in the alpine tundra," Dad warned Nick. "The mountain goat is one of the few large animals in the alpine tundra, and there aren't any mountain goats in the mountains where Grandma and Grandpa live. But if you are still interested by this summer I think we could convince Grandma and Grandpa to take us on a hike up past the tree line."

"Yippee! We're going on a *real* road trip!" Nick shouted.

**Research Assignment, pg. 63**

**Workbook pgs. 64-66**

Nature expresses a design of love and truth. It is prior to us, and it has been given to us by God as the setting for our life. Nature speaks to us of the Creator and his love for humanity. But it should also be stressed that it is contrary to authentic development to view nature as something more important than the human person.

— Benedict XVI, *Caritas in Veritate*

# 12
## Chapter

## Stewards of Creation

**Natural resources** are materials in the natural world that are useful to human beings.

Chemicals derived from the leaves of the Madagascar periwinkle have been very successful in treating certain types of cancer, including childhood leukemia.

## 12.1 Stewards of Creation

"During the last few weeks, we've studied seven biomes: the tundra, the boreal forest, the temperate forest, the rainforest, the savanna, the prairie, and the desert," Dad began. "In each biome we saw examples of God's loving care, which provides creatures with what they need to survive even in extreme climates. We also learned how plants and animals depend on each other, and we became familiar with the balance that exists within each biome.

"But there is another aspect to biomes that we haven't discussed up to this point. Biomes do more than provide habitats for different species of plants and animals. Each one also contains marvelous gifts from God, gifts that allow human beings to flourish on Earth.

"For instance," Dad said, "swamps and bogs play the important role of cleaning and replenishing groundwater supplies. The tundra is a rich source of oil, and thanks to its diversity of species, the rainforest is a source of many medicinal substances, not to mention delicious foods like chocolate and coffee. Solar power plants in the desert convert the intense desert sunshine into electrical energy to power our homes and cities."

"Termite mounds in the savanna have given architects ideas for new ways to keep buildings cool," Mike recalled.

"The bread we eat is made from wheat raised in the fertile soils of the temperate grassland," Dani contributed, "and our house is probably built out of timber from the boreal forest."

"Great examples!" Dad said. "Fresh water, timber, fertile soil, medicines, fuels—these are just a few examples of the many gifts God gives us through the natural world. Such gifts are called **natural resources**, and they are the next topic we are going to study."

"To begin," Mom said, "I want you each to name your favorite present of all time."

"Definitely my bike," Mike grinned.

"Rollerblades!" Nick shouted.

"Piano lessons," Dani said.

"My neenon crayons," declared five-year-old Christie. (Although Christie still had trouble pronouncing the word "neon," the refrigerator was covered with her colorful masterpieces.)

"I'm not surprised you chose those gifts," Mom smiled. "They have certainly given you many hours of enjoyment. But notice that they have required work and care on your part. For instance, Mike checks the tires on his bike before he goes for a ride, and he oils the chain regularly. To get any enjoyment out of her piano lessons, Dani has to spend time practicing every day. Nick has discovered that rollerblades don't roll when gum is stuck between the wheels, and Christie's crayons are neat and tidy because she is careful not to break them. Without care, all of these gifts would be broken or useless.

"The gift of natural resources must be cared for in the same way. When God created the world He made man the steward, or caretaker, of Creation. This means that the natural resources in Creation are meant to be used by man, but it also means that man has a responsibility to care for Creation."

"Do you mean the way farmers care for the soil?" Mike asked. "When we visited Josh and Hanna, their dad told us that farmers are very careful to plant the right crops at the right times to keep

the soil healthy. For instance, he said that if a farmer plants the same type of crop in a field year after year, the essential nutrients in the soil will be used up and plants won't grow on it anymore, at least not very well. To keep this from happening, farmers rotate the crops in their fields, and return nutrients to the soil by fertilizing it."

"That is an excellent example," Mom praised. "The fishing industry is another good example. Fish and seafood are staple foods in many parts of the world, and millions of people make their living by catching or raising fish. Just as farmers care for the soil, wise fishermen protect and care for the fish. With huge nets and special equipment, fishermen today can catch many fish at a time, but those who are wise are careful to leave some fish behind. These fish can then lay eggs that will hatch to fill the ocean with young fish. In a few years, the fishermen can return to catch a new generation of fish to feed more hungry people.

"Do you notice something in both of these examples?" Mom continued. "Fertile soil and abundant fish are important natural resources, but it takes wisdom and skill to care for them and put them to good use. Without human intelligence, these natural resources would not be the major food sources that they are. In fact, human intelligence itself could be thought of as a type of natural resource—the most important one, which allows us to make use of all the others.

"Of course, human beings are not 'resources' to be used the way we use soil and fish," Mom added. "But when God created human beings, He called us to use our intellects and our skills to care for Creation. When we act as stewards of the Earth by using natural resources wisely, we are acting as images of God the Creator, who cares for everything He has made. In the next few days, we will learn about particular situations in which man's intelligence can go to work to care for the Earth and provide for the human family," Mom concluded.

## 12.2 Life-Giving Water

The little country of Israel is one of the world's most productive agricultural regions. Grapes, cotton, grapefruit, avocados—all these crops and more are grown by Israeli farmers. Surprisingly, most of Israel is naturally a desert. Although the northern part of the country has an average precipitation of about 40 inches per year, areas farther south are much drier. In fact, southern Israel is as dry as the driest deserts in the United States, with an average precipitation of only two inches per year.

Even though southern Israel receives very little precipitation, there are many fertile farms there. Instead of relying on rainfall, Israeli farmers provide their plants with water through **irrigation**. Irrigation includes all the ways that man uses God's gift of water wisely by redistributing it to the places that need it most and by storing extra water for times of drought.

Since the northern half of Israel has plenty of water and the southern half of the country has scarcely any, Israel has built an incredible system of pipes, tunnels, and canals that transport life-giving water for hundreds of miles, from the Sea of Galilee in the north to the deserts in the south. Some of the pipes in the irrigation system are 10 feet tall! Israel has also built several large reservoirs, which are man-made bodies of water about the size of a lake. During the fall and winter, when Israel receives most of its rainfall, these reservoirs fill up with extra water. The reservoirs then release the water to farms and cities during the dry season.

Lake Powell, the second-largest man-made reservoir in the United States

**Irrigation** means using God's gift of water wisely by redistributing it to where it is needed most and by storing extra water for times of drought.

An aqueduct transporting water to Los Angeles, California

Flood Irrigation

Sprinkler Irrigation

Drip Irrigation

Wells and pumps are another source of water for farmers in Israel and other parts of the world. The water in wells and pumps comes from underground sources of water. Did you know that there is a lot of water underground? When water soaks into the ground after it rains, some of it is absorbed by thirsty plants, but much of it collects in **aquifers**. An aquifer is an underground source of water, and it is usually rock, gravel, or sand that is full of water. The water in an aquifer collects between the grains of sand and gravel, or in holes and cracks in solid rock. Wells and pumps are ways of extracting underground water from aquifers and bringing it to the surface.

Even though Israeli farmers receive water from northern Israel and from underground sources, they still would not have enough if they didn't know how to conserve water. To **conserve** a natural resource means to not use it or to take extra care not to waste any of it. For instance, people can conserve water during a drought by taking shorter showers, being careful not to leave the water running, and cutting back on how much they water their yard.

In Israel, many farmers conserve water by using drip irrigation to water their farms. A drip irrigation system delivers water directly to the roots of plants by channeling water through pipes or hoses. Farmers who use drip irrigation do not need as much water as they would if they irrigated by flooding their fields or by using sprinkler systems, because drip irrigation does not waste water on the bare dirt between plants. Also, very little water evaporates as it travels through the field, because the water is transported in pipes or hoses.

## 12.3 Caring for Forests

Workbook pg. 67

Do you remember learning about the importance of forest fires in the boreal forest? Even though severe forest fires might leave nothing behind but ash and dead tree trunks, they can actually prepare the way for new growth and healthier forests. For instance, jack pine cones do not release their seeds unless they are heated in a forest fire. When the seeds sprout, they grow and flourish because of the direct sunshine and nutritious ash that fills a newly-burned forest. Fire can also help a forest by cleansing it of diseases and insect pests.

Even though fire is good for many types of forest, especially boreal forests, there are also cases in which fire is harmful. For instance, some towns and neighborhoods are built near forests, and many people have homes in the forest itself. When a forest fire occurs in one of these forests, it can destroy the homes and possessions of many families. This is why forest fire fighters keep a sharp lookout for forest fires, and risk their lives to protect homes and towns.

Severe fires can also be harmful to the forest itself. Not all trees are like jack pines, with seeds that are tough enough to survive a fire. If a forest is not made up of trees like the jack pine, it has a hard time growing back after a severe fire. It is better if these forests only experience moderate fires, which kill some of the trees, but not all of them. The trees that are left are called **seed trees**, because they produce fresh seeds to fill the forest with a new generation of healthy trees.

So what makes some fires more severe than others? The answer is fuel. Severe fires usually occur in unhealthy forests that are full of dry leaves and dead trees. One way forests can become unhealthy is if human beings suppress, or prevent,

An **aquifer** is an underground source of water, and it is usually rock, gravel, or sand that is full of water.

To **conserve** a natural resource means to not use it or to take extra care not to waste any of it.

**Seed trees** are trees that are left after a forest fire. Seed trees provide the seeds for a new generation of trees.

**Foresters** are people who study, work in, and care for forests.

**Controlled burns** are small fires that foresters know they will be able to control. Foresters use controlled burns to keep forests healthy.

In nature, the believer recognizes the wonderful result of God's creative activity, which we may use responsibly to satisfy our legitimate needs, material or otherwise, while respecting the intrinsic balance of creation.
—Pope Benedict XVI, *Caritas in Veritate*

every single fire. When moderate fires are not allowed to burn the dry fuel and smaller trees, the forest becomes over-crowded. In crowded forests, there are not enough sunshine, water, and nutrients to go around, so the trees do not have the energy to fight off diseases and insect pests. Many of the trees in these crowded forests die, and the forest becomes even more cluttered with dry wood—the perfect fuel for a severe forest fire.

There are several ways that **foresters** can prevent severe fires. One way is to allow small and moderate fires to burn instead of putting them out right away. These fires will kill some of the trees, but the rest of the trees will have more space and more nutrients. A forest that is burned regularly by moderate fires remains healthy and strong, instead of becoming over-crowded and filling up with fuel for a severe fire. This is the natural way that God designed for temperate forests to remain healthy.

Allowing moderate fires to burn is not always the best way to care for a forest. If a moderate fire gets out of control, it can threaten homes and neighborhoods, or turn into a severe fire by spreading to an area that is full of dry fuel. Foresters sometimes solve these problems by starting small fires that they know they will be able to control. These small fires are called **controlled burns**.

Foresters can also keep a forest from becoming too crowded by harvesting, or cutting down, some of the trees. Trees that would otherwise die from over-crowding are cut down before they become fuel for severe fires. When some of the trees in a forest are harvested, the whole forest becomes healthier, because the

trees that are left no longer have to compete for space and nutrients. Careful harvesting can keep forests from filling up with too much dry fuel, which is a good way to prevent severe fires.

When foresters harvest the trees in a forest, it is good for the forest, and it is also helpful for human beings. Most of the trees that are harvested from the forest are turned into paper or lumber. The paper is used to make things like cardboard boxes, newspapers, and books, and the lumber is used to build houses, furniture, sailboats, pianos, pencils, and much more. What a useful way to save trees from being turned into charred stumps of wood by a severe forest fire! Careful harvesting of trees is an example of how man can use natural resources in ways that both protect God's Creation and benefit mankind.

## 12.4   Renewable and Non-Renewable Resources

Trees and fish are **renewable resources**, which means they can naturally replace themselves as they are used. Water and soil are also renewable resources. In most countries, these resources will never run out, so long as they are properly cared for.

Some natural resources are non-renewable, which means that there is only a certain amount on Earth, and if this is used up there is no way to get more. Coal, **petroleum**, and minerals such as iron, gold, and aluminum are **non-renewable resources**.

Do you think we should be worried by the thought of running out of non-renewable resources? Not at all! Many non-renewable resources, such as iron, gold, and aluminum, can be **recycled**, or used over and over. When we have sipped the last drop out of a can of soda, we can bring the aluminum can to a recycling center. The aluminum can will then be melted down and turned into a new can or other aluminum object.

**Renewable resources** are those that can naturally replace themselves as they are used.

**Petroleum** (puh-TROH-lee-uhm), or crude oil, is used to make gasoline, diesel fuel, propane, jet fuel, kerosene, and other products.

**Non-renewable resources** are those that cannot replace themselves.

To **recycle** means to use something over and over.

Besides learning how to recycle non-renewable resources such as iron, aluminum, and copper, human beings are also discovering how to use them more efficiently. For example, a single optical fiber (a hair-thin strand of ultra-pure glass) can carry as many telephone conversations as hundreds of copper wires. So if we started running out of copper, we could replace all the copper that we use for telephone wires with a much smaller amount of glass.

Of course, some non-renewable resources cannot be recycled. Coal and petroleum are important resources that cannot be reused. When your car is out of gas, it is no good trying to capture the used-up gas and put it back in the tank! No, the only solution is to go back to the gas station and fill up the tank with more gas.

Coal and petroleum are very important in our world today because they are two of our main sources of energy. A lot of our electricity is produced in coal-burning electrical power plants, and petroleum is the fuel that runs our cars and airplanes, heats our homes, and drives many machines.

Even though coal and petroleum are non-renewable resources, we should not be afraid of running out of them. Using the "natural resource" of our God-given intelligence, we have discovered many **alternative sources of energy** that we can use to replace coal and petroleum. For instance, new types of cars have been invented that run on electricity instead of on gasoline. Instead of producing electricity by burning coal, we can produce it through the energy of sunlight, the energy of moving water or wind, or the energy of nuclear reactions. Biomass such as wood, garbage, and alcohol fuels can also be burned to produce energy. A lot of our electricity is already produced through these methods, which are renewable or inexhaustible.

Do you notice a pattern here? Natural resources like water, forests, and oil are important, but the "natural resource" of man's

Major **alternative sources of energy**:
- Solar power
- Hydropower
- Wind power
- Biomass power
- Geothermal power
- Nuclear power

intelligence is even more important. Thanks to the gift of human intelligence, farmers in Israel are able to grow crops on land that was once a desert. By harvesting trees wisely, foresters are able to protect forests and homes from severe fires while also providing the human family with lumber and paper. Researchers are finding new supplies of non-renewable fuels and have discovered many renewable sources of energy. When man uses his intelligence to be a wise steward of God's Creation, he discovers that the world abounds with enough natural resources for all God's children.

Workbook pg. 68

**Selective breeding** is the process of breeding plants and animals to give them more desirable traits.

## 12.5 Growing Food for All God's Children

One way that man has used his intelligence to provide for the human family is by improving the different varieties of plants and animals through **selective breeding**. For example, Israeli dairy farmers developed the Israeli-Holstein breed by crossing native, drought-tolerant Damascus cows with Dutch bulls and Holstein-Friesian bulls, two breeds that are known for the abundance of milk they produce. The result? Dairy cows that produce incredible amounts of milk and are also well-adapted to the Middle Eastern climate.

Fruits, grains, flowers, and vegetables can also be improved by selective breeding. For instance, the American botanist Luther Burbank developed the Russet Burbank potato, which became one of the most common potatoes in the United States because of its large size, storability, and resistance to disease. The Russet Burbank potato saved thousands of lives when it proved to be resistant to the Irish potato blight that caused the Irish Potato Famine (1845-1852). Through selective breeding, Burbank developed over 800 new varieties of fruits, vegetables, nuts, grains, and flowers, including a spineless cactus to be used for animal feed in deserts.

Some people worry that the Earth might not have enough natural resources for the growing population of the world. But we know that man's intelligence is the greatest natural resource. There can't be too many people—there can only be too little ingenuity, creativity, and generosity.

Today, scientists can breed plants even more selectively by using the latest biological technology. Scientists have learned how to isolate the **genes**—the bits of coded information (DNA) that tell a plant how to grow—for desirable traits such as resistance to drought or disease. They can then insert the gene into a different plant, making it resistant to drought as well. New varieties of plants that are produced this way are called **genetically modified crops**.

Of course, genetically modified crops have to be carefully tested before they can be grown outside the laboratory. Scientists would not want to accidentally grow plants that are unhealthy for people to eat! So long as genetically modified foods are carefully tested, they can be a great blessing to mankind. Scientists have already developed a variety of corn that makes its own **pesticide** against insects such as the European corn borer. Researchers have also developed a "golden" rice that contains large quantities of vitamin A, an essential vitamin for good eyesight. Normal rice contains very little vitamin A. In poorer parts of the world, many people go blind because they cannot afford any food but rice. If countries begin to grow the new "golden" rice, these people will be able to get vitamin A more easily.

PLANTING RICE

GOLDEN RICE (LEFT) COMPARED TO WHITE RICE (RIGHT)

## 12.6 Conservation

"Water, soil, forests, sunshine, coal and oil, minerals such as gold, iron, and copper—all of these are natural resources. Natural resources are the gifts God has given us in Creation to care for and to use wisely. Now, what is the first thing you do when you receive a gift?"

"Say 'thank you'?" Mike suggested.

"Exactly," Dad agreed. "This is true for birthday presents, and it is also true for God's gift of natural resources. Our first response to God's gift of Creation is one of gratitude and awe, and this gratitude leads us to respect and care for Creation.

"When we study how to care for Creation and use natural resources wisely, we are studying **conservation**, which is part of ecology," Dad continued. "At the beginning of our study of biomes, you learned that the word ecology comes from the Greek word for house. This is very fitting, because in a certain sense, the natural world is the home that God created for the human family.

"The way we take care of Creation is similar to the way we take care of our home. The reason we have rules like, 'Draw on paper, not on the walls,' and 'Wipe your feet before you come inside,' is because we are grateful for the house God has given us, and we want to take care of it. We are also careful to conserve our family's resources. For instance, we turn the lights off when we leave a room to avoid wasting electricity, and we don't leave the water running when we aren't using it. We are careful not to waste food because we want to show our gratitude to God, who gives us what we need to stay strong and healthy.

**Genes** are the bits of coded information (DNA) that tell a plant how to grow.

**Genetically modified crops** are plants whose genes have been changed through biological technology.

**Pesticides** are chemicals that kill insect pests.

**Conservation** is the study of how to care for Creation and use natural resources wisely.

"Of course, part of taking care of our home is taking care of the people who live in it. When Nick fell off the swing the other day and scraped his knee, Mom used all her skill and lots of cotton balls, hydrogen peroxide, and bandages to care for the wound. By eating nutritious foods and getting plenty of sleep and exercise, we are caring for the gift of health that God gave to us. When you study science, history, and other subjects, you are being wise stewards of the intelligence that God has given you, and you are showing your gratitude for the gift of education.

"Now, in every home there are many things that need care, but notice whom we care for first. When Nick fell off the swing, did Mom tell him, 'Hold on a minute—I need to finish vacuuming the living room'? Of course not! She dropped everything she was doing to take care of Nick.

"What if we were snowed in for a week and couldn't get to the store to buy groceries? If we started running low on food, would we give the last bit of food to our dog? Definitely not! We would give it to you children. We are grateful for our house and our dog, but people are more important than animals and things. This is because every human being is created in the image and likeness of God, and is called to live with God forever in Heaven."

Dad looked at Mom and smiled before he continued. "And that's why your Mom and I are so excited—we have learned that God is blessing us with a new member of our family."

Mike grinned as he realized what Dad meant. "You mean . . ." he began.

Dani interrupted him with an excited shriek: "Mom, are you going to have a baby?" When Mom nodded, Dani jumped up and hugged Christie, whirling her around the living room.

> The way humanity treats the environment influences the way it treats itself, and vice versa. . . . It is contradictory to insist that future generations respect the natural environment when our educational systems and laws do not help them to respect themselves.
> —Pope Benedict XVI, *Caritas in Veritate*

Nick took the opportunity to turn several cartwheels, narrowly missing the edge of the fireplace. "When the baby is born," he declared breathlessly, "I'm gonna share my rollerblades with him and teach him to do handstands."

"We don't know yet if the baby is a boy or a girl," Dad reminded him, "but I am glad you are so ready to be a good big brother. Like all precious gifts, the new baby God is entrusting to us will need lots of loving care. In the months before the baby is born, we'll get the baby clothes out of the garage and dust off the crib. Mom will make sure she eats the right foods to keep the baby healthy, and we will help her with the housework when she feels tired."

"Then when the baby is born," Mom continued, "there will be lots of chances to rock the baby to sleep, to help give the baby a bath, and to entertain the baby. When the baby starts to crawl and walk, you can help rearrange the furniture so there aren't as many sharp corners for him—or her—to fall against."

"If the baby is a boy, can he sleep in our room?" Mike asked eagerly.

"The baby will sleep in our room at first," Mom said, "but when the baby is older he or she will get to join the 'big kids' in the boys' room or the girls' room. There's plenty of space if we stack the bunk beds."

"I want to sleep in the top bunk!" Nick shouted.

"We'll discuss that when the time comes," Mom smiled. "Right now, let's celebrate the gift of new life that God has given our family. Who wants a slice of homemade chocolate cheesecake?"

> Openness to life is at the center of true development. . . . The acceptance of life strengthens moral fiber and makes people capable of mutual help.
> —Pope Benedict XVI, *Caritas in Veritate*

 Workbook pg. 69

 Unit 2 Test pgs. 70-71

UNIT 3

# 13
Chapter

# Constellations and Star-Gazing

*If the stars should appear but one night every thousand years how man would marvel and stare.*
— Ralph Waldo Emerson

## 13.1 Constellations and Star-Gazing

On an evening in March, soon after completing their virtual tour of the biomes, Dad and Mike met in the living room. Mike carried his textbook under his arm, and Dad bore a stack of star-gazing resources. "Did Mom tell you what we'll be studying now that we have completed our tour of the biomes?" Dad asked his son.

"She said you're going to teach me all about astronomy!" Mike said. "Did you really get to look through a giant telescope when you studied astronomy in college?"

"I sure did. It was amazing to be able to see the rings of Saturn through a telescope larger than myself," Dad said. "But most of what I learned about astronomy didn't require a telescope. Did Mom explain to you what astronomy is?"

"Yes," Mike affirmed. "**Astronomy** is the study of the stars, the planets, and all the other objects in outer space. Mom said that when we study astronomy we aren't just trying to memorize the positions of the stars or what the Sun is made of. Instead we are trying to understand the scientific laws that explain why the stars and planets are the way they are."

**Astronomy** is the scientific study of the stars, the planets, and all the other objects in outer space.

Constellations >>

"And that's the reason we can study astronomy from the kitchen table and in our own backyard," Dad said. "Since astronomy is more than star-gazing and the memorization of facts, understanding astronomy only requires a good brain and an active imagination."

"But we are going to do some star-gazing, right?" Mike asked eagerly. "Mom said I'll get to stay up late sometimes to look at the stars with you."

"Yes, we'll begin our study of astronomy by looking at the stars from our own backyard and learning how to find our way around the night sky. This makes sense, because everything we know about astronomy was originally discovered by scientists who pondered what they could see in the night sky. And besides, 'the heavens declare the glory of God,' and it would be a pity to study astronomy without taking time to appreciate the grandeur of the heavens."

"It's too bad it's so cold out at this time of year," Mike said. "It'd be more fun to go star-gazing in the summer."

"I'm sure we'll do a lot of star-gazing this summer," Dad said. "But I think it's worthwhile to study astronomy at this time of year, because most people find the winter constellations easier to recognize than the summer constellations."

"A **constellation** is a group of stars that makes a picture, right?" Mike asked.

> A **constellation** is a group of stars that makes a picture. The word also refers to the region of the sky around the group of stars.

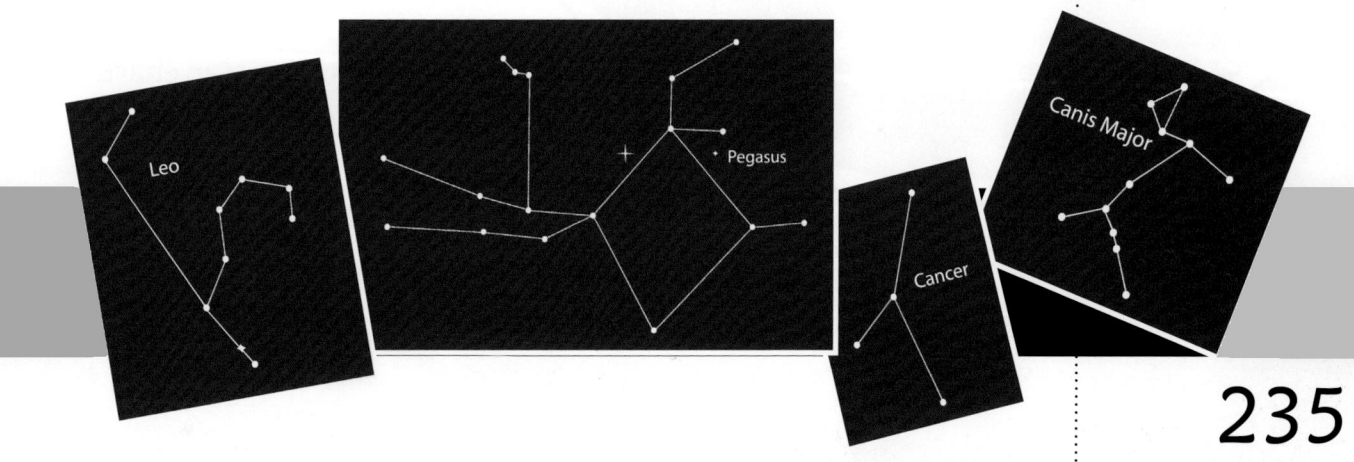

**Orion** (oh-RYE-un)

Seasonal star charts can be downloaded for free from websites such as www.skymaps.com.

"That's how we'll be using the word," Dad affirmed. "The Little Dipper is a good example of a constellation. So is **Orion**, which is the constellation that portrays Orion the Hunter battling off a wild bull with a shield and a club. Orion is probably the easiest constellation to recognize because of the three stars that form his belt."

"Can I go outside right now and see if I can find him?" Mike asked eagerly, already struggling into his coat.

"First let me show you how to use a star chart. Then we can use the chart to locate Orion."

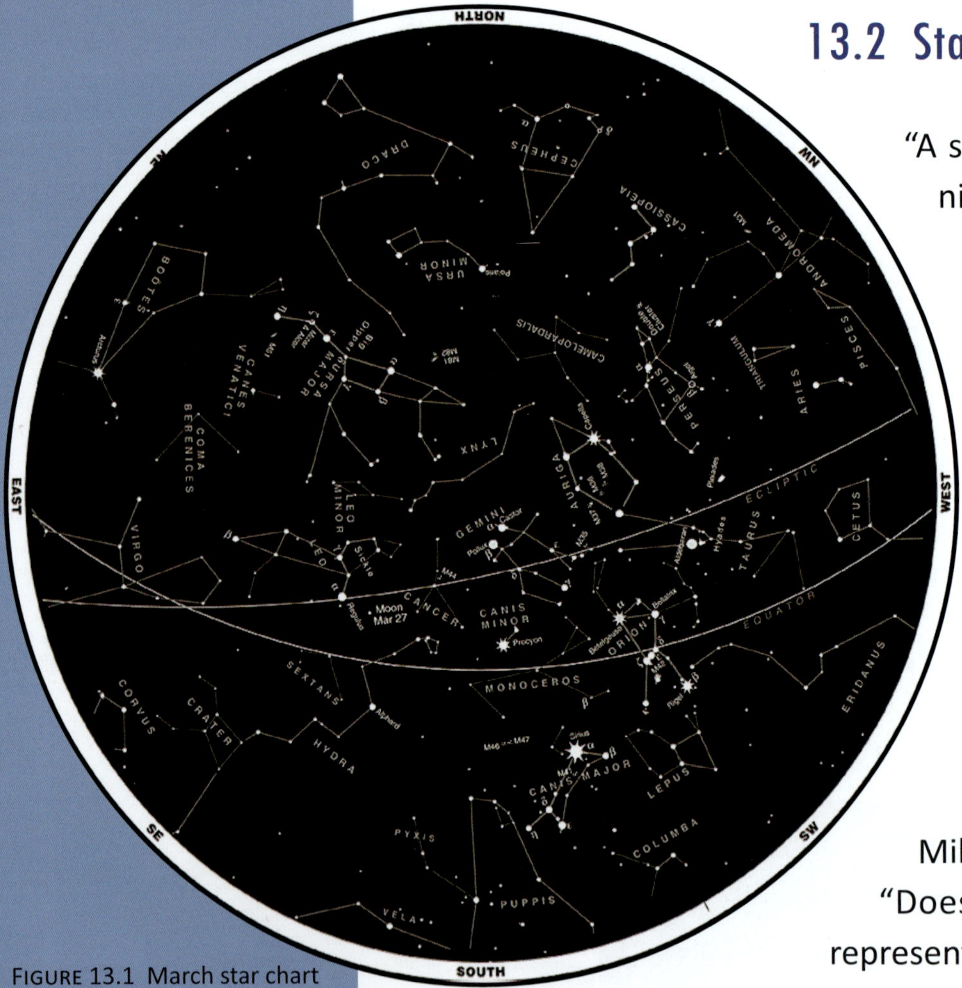

## 13.2 Star Charts

"A star chart is a map of the night sky," Dad explained. "Star charts are often drawn in the shape of an oval or a circle, because the night sky looks like a round dome when we look up at it. This star chart shows what the night sky looks like at around 8 PM in March if you are in the northern hemisphere."

Mike studied the star chart. "Does the edge of the chart represent the horizon?" he asked.

"Yes," Dad said. "And the center of the chart represents the part of the sky directly above your head."

FIGURE 13.1 March star chart

Figure 13.2 Star charts should be held upside down.

"The horizons are labelled wrong on this star chart," Mike commented. "The chart has north on the top and south on the bottom, but east and west are switched."

"I'm glad you noticed that!" Dad said. "The chart looks like it was labelled incorrectly, but that's because it's meant to be held upside down.

"Try this: face south, and then hold the star chart so the southern horizon is at the bottom of the chart. Now raise the star chart above your head. Are the horizons labelled correctly now?"

"Yes! The southern horizon is in front of me, the eastern horizon is on the left, and the western horizon is on the right," Mike said. "I guess the chart looks different because it is a chart of the sky, not the Earth."

"Exactly. Now see if you can find Orion on the chart. It is a little bit south of the center."

"Found it!" Mike said.

"When we go outside, which direction should we face to see Orion in front of us?" Dad asked.

"In the star chart, Orion is closest to the south-western horizon," Mike observed. "Should we face south and a little west?"

"You've got it!" Dad praised. "Now let's go outside and study the stars!"

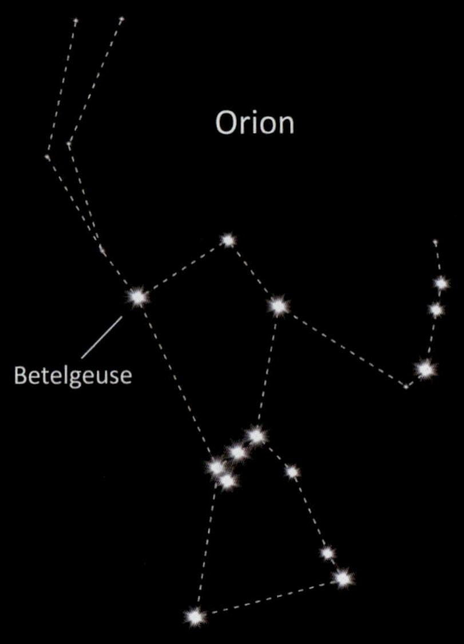

**Orion**

Betelgeuse

FIGURE 13.3 According to Greek myth, Orion's shield was made from the pelt of a lion.

John R. Foster/Photo Researchers

Do you know how to find Orion in the night sky? If not, imitate Mike and his dad by using a star chart to find it in the sky. Seasonal star charts can be downloaded for free from websites such as www.skymaps.com.

## 13.3 Finding Constellations in the Night Sky

"I found him!" Mike exclaimed. "There's Orion's belt, and the stars for his shoulders and knees. The reason I couldn't find him at first is that he is sort of crooked in the sky."

"Good job," Dad said. "Do you notice the bright star a little below Orion's belt? That's his dagger. His shield and club are harder to see, but they are located to the right and left of his shoulders."

"I think I see them," Mike said. "But where is Orion's head? Is he the Headless Hunter?" he joked.

"His head resides in your imagination," Dad chuckled. "That way you can draw his features however you want."

"And his legs and feet are in my imagination, too, I suppose," Mike said.

"Exactly! You're really getting the hang of this!" Dad laughed. "It's true, a lot of the constellations have missing heads, arms, legs, and more. With a good imagination, though, you can start to see the constellations as actual pictures instead of just jumbles of stars.

"Now look at the bright, reddish star at the top left corner of the constellation Orion," Dad continued. "That is a very famous star called **Betelgeuse**, and it marks one of Orion's shoulders. It's important that you memorize the names of a few major stars because that will help you find your way around the sky more easily. Besides Betelgeuse, you should find and memorize the stars named **Sirius** and **Procyon**. They are the main stars in **Canis Major** and **Canis Minor**."

"Are those constellations, too?"

"Yes. Canis Major and Canis Minor are Orion's hunting dogs. Their names mean 'big dog' and 'little dog.'"

"How funny!" Mike said. "Where are they?"

"Their main stars are so bright that I could just point them out to you, but let's see if you can find them by using your star chart."

Mike searched the star chart for Sirius and Procyon. "There's Sirius in Canis Major," he said. "It's a little behind and below Orion. And Procyon is a little behind and above Orion, in Canis Minor."

"Yes, Orion's dogs are following him to the hunt," Dad said. "Canis Major is jumping up from the ground towards Orion. Sirius is a bright star near his neck, like a shiny dog tag. Procyon is the brightest star in Canis Minor."

"There's only one other star in Canis Minor," Mike noted. "The 'little dog' isn't much more than a tail! I'm sure he's doing his best to help Orion, though."

"Now your imagination is working!" Dad said. "Let's see if you can find Sirius and Procyon in the sky."

Mike studied the sky behind Orion. "I see two bright stars really far behind Orion. But they aren't close enough to Orion to be Sirius and Procyon, are they?"

"Those stars are indeed Sirius and Procyon," Dad said. "What looks like a small distance on a chart is often much bigger in the sky. You can use your fist to estimate distances between constellations. Make a fist and hold your arm out straight. Now point your arm towards Orion and tell me how many fists you can fit between his belt and his shoulders."

"About one fist," Mike said.

**Betelgeuse** (BET-el-jooz)

**Sirius** (SEER-ee-us)

**Procyon** (PRO-see-on)

**Canis Major** (KAY-niss MAY-jer)

**Canis Minor** (KAY-niss MY-ner)

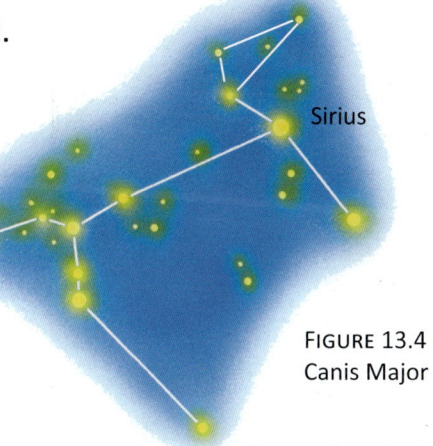

FIGURE 13.4 Canis Major

Sirius is the brightest star visible in the Northern Hemisphere.

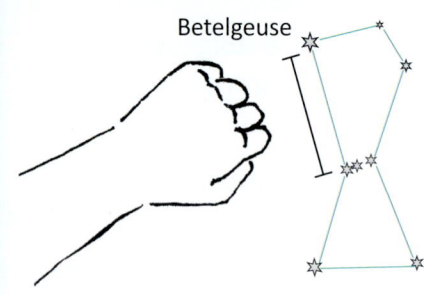

FIGURE 13.5 The distance between Betelgeuse and Orion's belt is one "fist-length."

"So let's look at the star chart again," Dad said. "If the distance between Betelgeuse and Orion's belt equals one fist, how many 'fist-lengths' can you fit between Betelgeuse and Procyon, and between Betelgeuse and Sirius?"

Mike estimated the distances on the star chart. "Three fists from Betelgeuse to Sirius," he said. "And three fists from Betelgeuse to Procyon."

Looking up from the star chart, Mike held his arms out straight and used his fists to measure the distances between Betelgeuse and Sirius. "You're right!" he exclaimed. "Sirius is exactly as far from Betelgeuse as the star chart says it is. And Procyon is, too."

"You figured that out nicely!" Dad praised. "I think you'll find your fists very useful for estimating distances in the sky. In fact, let's use them again to find Taurus."

"What's that?" Mike asked.

"**Taurus** is the wild bull that Orion is battling against. In the sky, Taurus has lowered his horns and is preparing to charge at Orion. The main stars you should look for in this constellation are

**Taurus** (TOR-us)

**Aldebaran** (al-DEB-uh-ran)

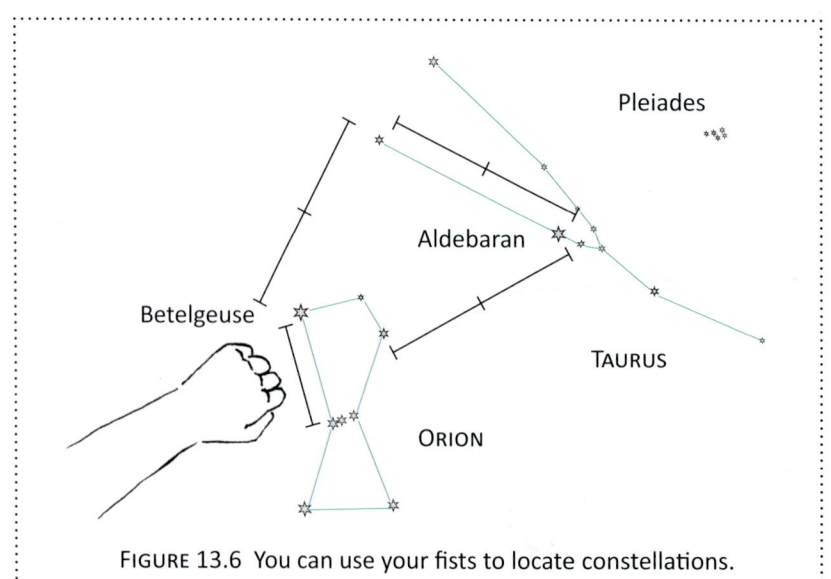

FIGURE 13.6 You can use your fists to locate constellations.

**Aldebaran**, which is Taurus' eye, and the two stars marking the sharp points of Taurus' horns. These three stars form a large triangle above and in front of Orion."

Mike gazed at the sky. "I don't see any triangle, but maybe I should use my fists to estimate the distances."

Looking at the star chart, he calculated, "I should be able to fit almost two fists between Betelgeuse and Taurus' lower horn, and two fists between Aldebaran and Orion's other shoulder. The distance between Aldebaran and the two horns is also about two fists, or maybe a little less."

Mike turned back to the sky, holding up his fists to measure the distances to Taurus. "I see it! It's just a lot bigger than I expected, so I couldn't find it at first. What should I find next?" Mike asked eagerly.

"It's too cold to stay out here much longer, but I'd like you to see the **Pleiades** before we go inside. They aren't very bright, but they are close together in the sky, and quite beautiful. You can see them about one and a half fists behind Aldebaran."

Mike held up his fists again until he found the Pleiades, a faint cluster of stars high in the sky. "Ta-daa!" he exclaimed. "Found them!"

"According to the ancient Greeks, the Pleiades are seven sisters who were turned into stars."

"I can only see six stars," Mike said. "Maybe the ancient Greeks had better eyesight than we do!"

"They certainly had less **light pollution** from street lights!" Dad laughed. "But they couldn't go inside and enjoy a cup of hot cocoa after star-gazing, as we can!"

"Yum!" Mike exclaimed. "Let's go!"

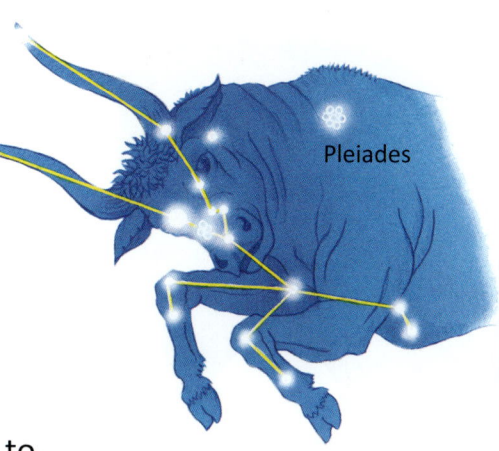

FIGURE 13.7 Taurus

**Pleiades**
(PLEE-ah-deez)

**Light pollution** is the glow from city lights that obscures our view of fainter stars.

Unit 3 Activity #1, pg. 73

FIGURE 13.8 Pleiades

## 13.4 Light-Years and AU's

"Last night we estimated the distances between constellations by using our fists," Dad said. "But is Betelgeuse *really* three fist-lengths away from Sirius?"

"You mean in outer space? Of course not!" Mike laughed. "Betelgeuse is actually hundreds of thousands of miles from Sirius. The distance between them just *looks* like three fists because Betelgeuse and Sirius are so far away."

"You're exactly right," Dad said. "Even Proxima Centauri, which is the closest star we know of, is 25 trillion miles away—that is 25 with 12 zeros after it."

"Whoa!" Mike exclaimed. "I can't even imagine a distance that huge!"

"But 25 trillion miles is just a hop, skip, and a jump in astronomy," Dad said. "Betelgeuse and Sirius are at least two and a half quadrillion miles (2,500,000,000,000,000 miles) from each other, and probably much more."

"That is definitely more than three fists!" Mike gasped. "Astronomers must be really good at math to deal with all these long numbers."

"You're right that astronomers are good at math," Dad said, "but they are also smart enough to make their job easier by using different units of distance."

"Units of distance?" Mike repeated. "You mean like inches and feet?"

"Exactly. In our daily lives we use units like centimeters, inches, feet, and miles, but astronomers have made up their own units to help them work with the huge distances between planets, stars, and galaxies. The two main units of distance in astronomy are the **light-year** and the **AU**, which is short for astronomical unit."

"'Light-year' sounds like it's a unit of time, not a unit of distance," Mike objected.

"I know," Dad said. "In fact, non-scientists are often confused by light-years, and think that a scientist is talking about time when he is actually talking about distance."

Mike silently resolved never to make this mistake. "So how far is a light-year?" he asked.

"A light-year is equal to about six trillion miles. The light-year gets its name because six trillion miles is the distance that light can travel in one year. Light moves faster than anything else in the universe, so it can travel pretty far in a year!"

"I didn't know it took time for light to travel from place to place," Mike said. "When I turn on a light it fills the whole room instantly."

"That's because light travels so quickly. It takes less than a billionth of a second for light to travel across your bedroom. A ray of light could travel around the globe 10 times in just one second."

"Gosh!" Mike exclaimed. "How long does it take light to travel from the Sun to the Earth?"

"About eight minutes," Dad said. "And that brings us to the second unit of distance that astronomers use: the AU, which stands for 'astronomical unit.' The AU is the distance between the Earth and the Sun, and it is equal to about 93 million miles."

"That sounds just as big as a light-year," Mike admitted. "These huge distances make my head spin."

---

A **light-year** is the distance light can travel in a year, about six trillion miles.

An **AU**, or **astronomical unit**, is the distance between the Earth and the Sun, about 93 million miles.

Proxima Centauri, the closest star to our Sun, is about four light-years distant. This means that it takes four years for the light from Proxima Centauri to reach the Earth.

"I know how you feel," Dad said, "but it is important to understand the difference between an AU and a light-year. An astronomical unit is a huge distance, but it is 63,000 times smaller than a light-year."

"More big numbers!" Mike groaned.

"Don't worry," Dad reassured Mike, "the most important facts to remember are that a light-year is the distance that light can travel in one year, and an AU is the distance between the Sun and the Earth."

"And a light-year is a unit of distance, not of time," Mike added.

"Right," Dad agreed. "If you can get that down, you will be well ahead of most people your age—or even my age."

## 13.5 Appearances versus Reality

A star chart is a map of what we can see in the night sky; it is not a map of where the stars really are. We said that a constellation is a group of stars that makes a picture. This is true, but often the individual stars in the constellation only *look* like they are near each other, and are quite far apart in reality. For instance, on a star chart Betelgeuse and Procyon look like they are the same distance from Sirius (Figure 13.1). In reality, Betelgeuse is at least 20 times farther from Sirius than Procyon is.

Consider the bird and airplane in Figure 13.9. In the photo, the airplane and the bird appear to be the same distance from the tip of the pine tree. But when the photo was taken, the airplane was actually thousands of feet from the pine tree, while the bird was only a few feet away. In the same way, the Betelgeuse-Sirius distance is longer than the Sirius-Procyon distance, even though the distances look the same on the star chart.

Do you notice something else about Figure 13.9? In the photo, the bird looks twice as large as the airplane, even though we know this is not the case in reality. The bird looks larger only because it is so much closer to us than the airplane is. It is the same with the stars. There are many different types of stars, and some are hundreds of times brighter than others. But a very bright star will look quite dim to us if it is far away. And a star that is quite dim can look extremely bright if it is close to us.

To understand this better, imagine standing outside on a dark night with a single candle. Down the road, your neighbors are roasting marshmallows over a bonfire. Because you have been studying astronomy lately, you notice that your candle looks much brighter than the neighbor's bonfire. Of course, you know that the bonfire is much brighter in reality, but the candle *seems* brighter because you are so much closer to it.

The same principle applies to the stars. The case of Sirius and Betelgeuse is a good example. In the sky, Sirius looks much brighter than Betelgeuse, but in reality, Sirius is just a candle held up in front of the Betelgeuse-bonfire. If Betelgeuse and Sirius were the same distance from Earth, we would see that Betelgeuse is more than 1400 times as bright as Sirius. But the Betelgeuse-bonfire is 50 times farther from the Earth than the Sirius-candle is, so the Sirius-candle looks much brighter.

FIGURE 13.9 The bird and the airplane appear to be the same distance from the tree.

FIGURE 13.10 A candle and a bonfire

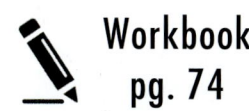

Workbook pg. 74

## 13.6 The Motion of the Stars

"It's 11 o'clock, Dad!" Mike whispered.

"Hmmm?" Dad grunted from the living-room couch. He and Mike were staying up late to look at the stars, but Dad had fallen asleep over his newspaper.

"Can we go star-gazing now?" Mike asked. "I have my coat on already."

"Okay, just give me a hand up from this couch," Dad said. "I was dreaming I landed on the Moon, and it was covered in clouds of soft whipped cream. It was sooo comfortable."

"You're joking," Mike said.

"Yes, I am," Dad agreed. "But the couch *was* very comfortable. You go on ahead, and I'll join you in the backyard as soon as I have my coat and boots on."

A few minutes later, Dad quietly shut the back door and walked over to where Mike was standing and gazing at the sky. "How's it going?" Dad asked. "Have you found Orion?"

"Yes, but it's in a different place!" Mike exclaimed. "Is that because we're star-gazing a few hours later than we usually do?"

"Yes," Dad affirmed. "Just as the Sun moves across the sky during the day, so the stars move across the sky during the night. Of course, the Sun doesn't actually move through the sky, does it?"

"No," Mike said. "The Sun looks like it's moving because the Earth is rotating on its **axis**. When our side of the globe is facing the Sun, it's day. When our side of the globe is facing away from the Sun, it's night. One **rotation** is the same as one day and night."

An **axis** is the imaginary center "line" around which an object rotates.

To **rotate** means to spin. One **rotation** of the Earth is the same as one day and night.

"Excellent!" Dad praised. "The stars seem to move across the sky at night for the same reason. Since the Earth is turning on its axis, we see different parts of the sky all night long."

"So if we go star-gazing at 5 AM we would see completely different stars from those we see at 8 PM?" Mike asked.

"A lot of the stars would be different, but not all of them," Dad said. "The North Star and the stars around the North Star are always visible to us."

"How come?" Mike asked. "Don't all the stars move across the sky together?"

"Well, the stars aren't really moving, remember. They only appear to move because the Earth is rotating under the sky. But you're right—the stars all 'move' across the sky together.

"Because we live between the North Pole and the Equator," Dad continued, "the stars don't move straight across the sky. Instead, they move across the sky in circles and curves, because they are traveling—I mean, they *seem* to be traveling—around the North Star. The North Star is the one star that stays in the same place in the sky all night long."

FIGURE 13.11 Only the stars on the dark side of the Earth are visible. The others are present, but cannot be seen because their light is drowned out by the light of the Sun.

FIGURE 13.12 This time-lapse photo of the night sky shows the path of the stars over the course of 17 minutes.

"Hey!" Mike exclaimed. "That must be why characters in books always use the North Star to find their way home when they are lost."

"That's right! Since the North Star stays in the same place all night, you can use it to figure out the directions of north, south, east, and west," Dad explained. "In fact, let me show you right now how to locate the North Star in the sky. We can wait until later to learn *why* the rotation of the Earth makes the stars seem to travel around the North Star."

## 13.7 Polaris, the North Star

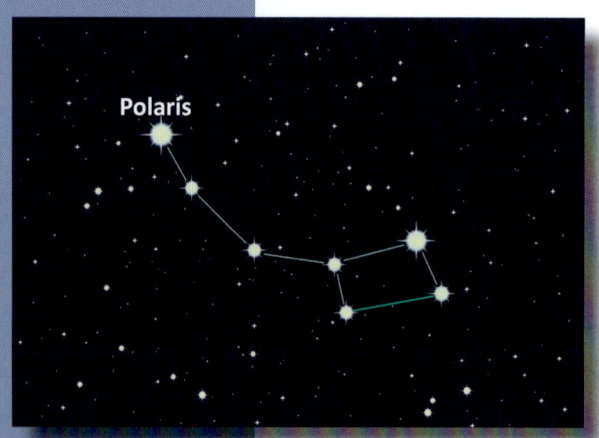

FIGURE 13.13 Ursa Minor (Little Dipper)

The North Star's official name is **Polaris**, and that is the name you will see on star charts. If you live north of the Equator, you can use Polaris to find the direction of true north.[1]

Often, people are surprised to learn that Polaris is not a bright star, but this is because they do not understand the real reason that Polaris is special. Polaris is not special for its brightness or beauty; it is special because it never moves, all night long. Because of this, Polaris is a fixed point in the sky that can guide travellers through the wilderness back to their homes.

**Polaris** (poh-LAIR-iss)

Polaris is part of the constellation Ursa Minor, which we usually call the Little Dipper. Ursa Minor means "Little Bear" in Latin. Polaris is positioned at the very tip of the Little Dipper's handle, which is also the very end of the Little Bear's unusually long tail.

The Little Dipper is not always easy to locate in the sky because its stars are not especially bright, so people often use the Big

---

[1] If you live south of the Equator, use the resources listed on page 74 in the workbook to learn how to find the direction of true south using the Southern Cross.

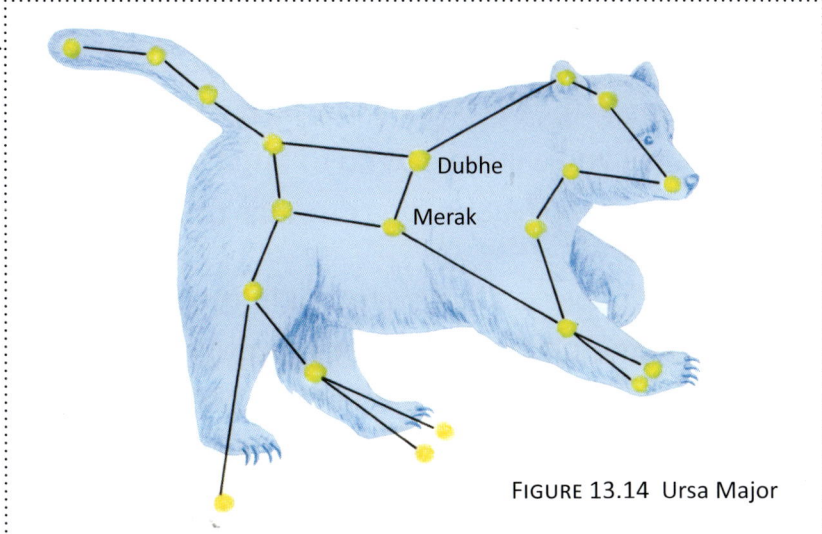

FIGURE 13.14 Ursa Major

Dipper to help them find Polaris. The Big Dipper is part of the constellation that star charts call Ursa Major, which means "Big Bear." The Big Dipper is about the size of Orion, and it can be seen in the northern part of the sky at any time of the year, all night long.[2] You might already know how to find the Big Dipper, but if not, you can use a star chart and your fists to find it.

Once you know how to find the Big Dipper, locating Polaris is simple because two of the stars in the Big Dipper point directly to Polaris. These two stars are called **Merak** and **Dubhe**, and you can see them in Figure 13.15. They are on the side of the Big Dipper opposite its handle. You should memorize their names

**Merak** (MER-ak)

**Dubhe** (DOO-bee)

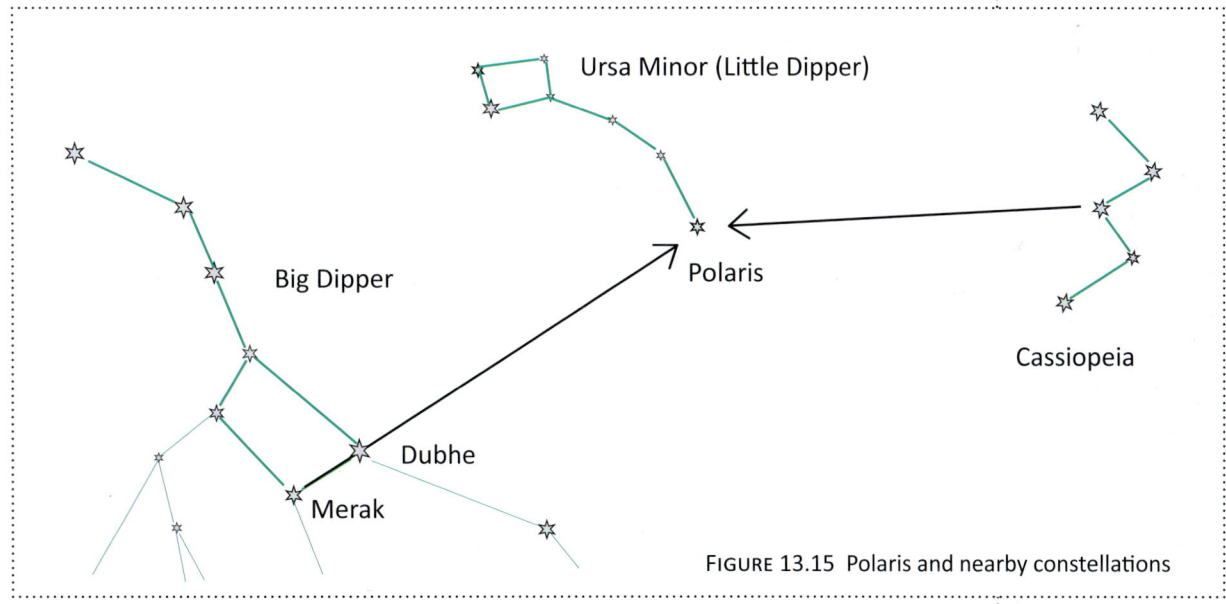

FIGURE 13.15 Polaris and nearby constellations

---

[2] Unless you live south of the Equator.

and their positions in the Big Dipper. The next time you go outside on a clear night, all you will have to do to find Polaris is follow the line between Merak and Dubhe for a distance of about three fists.

Although the Big Dipper is always present in the night sky, it does not stay in the same place. As it moves through the sky, the Big Dipper makes a wide circle around Polaris. This means that the Big Dipper is sometimes above Polaris and sometimes below it. It also means that it is sometimes right side up, sometimes upside down, and sometimes tilted on its side.

The constellation **Cassiopeia** can also be helpful for finding Polaris. Cassiopeia is about half or a third the size of the Big Dipper, and it is named after a famous queen in Greek mythology. As you can see in Figure 13.15, Cassiopeia is made of five stars arranged in a wide, flat "W" or "M." Polaris is always about three fists below the "M" of Cassiopeia (or above the "W").

Like the Big Dipper, Cassiopeia can be found in the northern sky all year round, at any time of night, so long as you live north of the Equator. Also like the Big Dipper, Cassiopeia changes its position and orientation during the course of the night because it travels around Polaris. Cassiopeia and the Big Dipper are always on opposite sides of Polaris, so if one of them is hidden behind trees or houses, you can use the other constellation to find Polaris.

**Cassiopeia**
(KAS-ee-oh-PEE-ah)

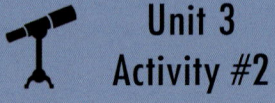
Unit 3
Activity #2

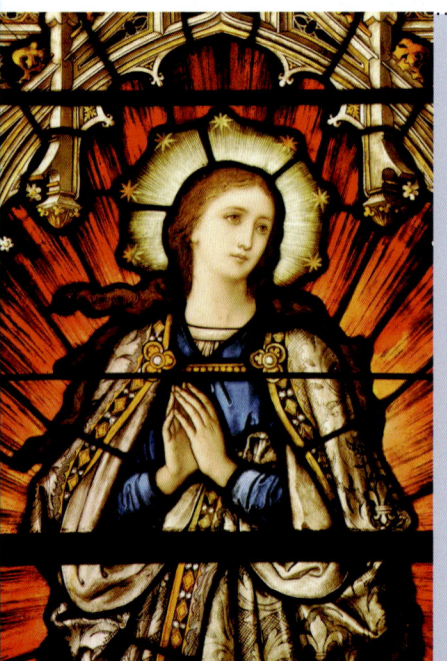

### Our Lady, Star of the Sea

The stars were so important for knowing directions at night on the ocean that sailors saw them as an image of Mary, and called Mary "Star of the Sea." Just as the stars would guide sailors safely home, so Mary will guide Christians safely to Heaven.

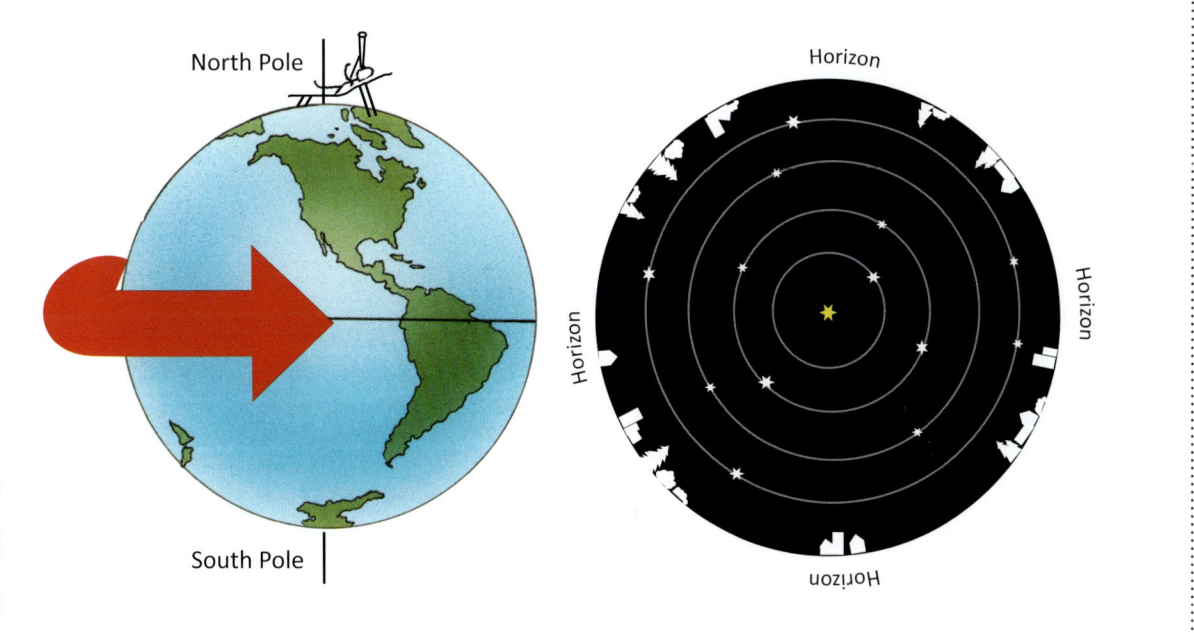

FIGURE 13.16 To a person at the North Pole, the stars appear to move around the sky in circles.

## 13.8 Understanding Rotation and Revolution

"So why do the stars travel around Polaris?" Mike asked. "I mean," he corrected himself, "why do they *look* like they are traveling around Polaris?"

"Good catch," Dad said. "You're right—the stars don't actually move, but they seem to move because the Earth is rotating on its axis. And the stars seem to travel around Polaris because Polaris is the star directly above the North Pole," Dad explained. "If we lived at the North Pole, Polaris would be the highest star in the sky. At the North Pole, the stars do not rise and set—instead they travel around the sky in a circle, with Polaris at the center.

"Living at the North Pole is a little like sitting at the center of a merry-go-round," Dad continued. "If you look straight up at the sky while the merry-go-round is turning, most of the sky will be rotating in circles around your head. The highest point of the sky won't move because it is directly above your head. At the North Pole, this highest point is Polaris, the North Star.

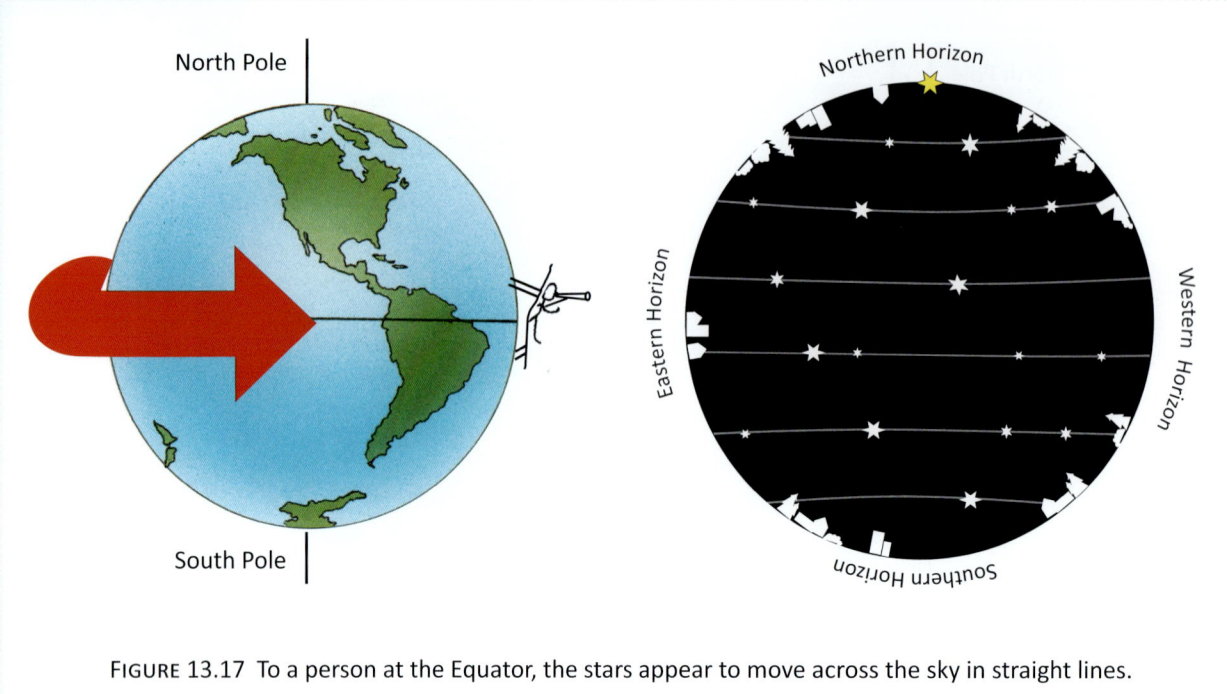

FIGURE 13.17 To a person at the Equator, the stars appear to move across the sky in straight lines.

"If you were star-gazing at the Equator," Dad continued, "Polaris would stay in one place near the northern horizon. The other stars would still be rotating around Polaris, but from your point of view they would seem to be moving straight across the sky. There's a picture in your science book that shows how this would look." (Figure 13.17)

Mike studied the picture for a while, and then Dad continued. "Since we live in North America, our house is located between the Equator and the North Pole. So we see Polaris several 'fists' up from the northern horizon, in between the horizon and the highest point of the sky. The stars near Polaris travel around it in small circles, and stars that are farther from Polaris move across the sky on slightly curved paths.

"As you can see in Figure 13.18," Dad said, "stars that are near Polaris are visible all night long because their 'circles' never dip below the horizon. Dubhe and Merak in the Big Dipper are examples of such stars."

"But stars like Betelgeuse and Sirius are farther away from Polaris,"

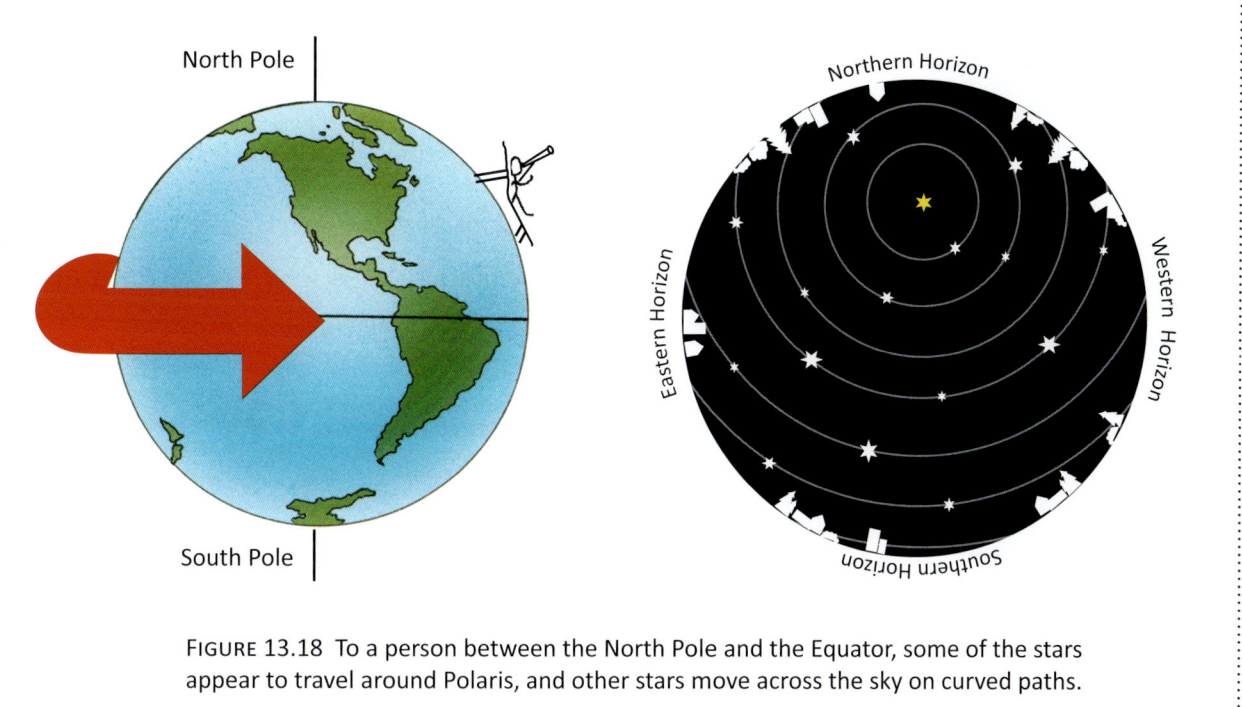

FIGURE 13.18  To a person between the North Pole and the Equator, some of the stars appear to travel around Polaris, and other stars move across the sky on curved paths.

Mike said, "so sometimes their 'circles' are below the horizon. That's why we couldn't see Sirius at 11 o'clock the other night! It had already set below the western horizon."

"Precisely," Dad said.

FIGURE 13.19  This time-lapse photo of the night sky shows the path of the stars over the course of several hours.

Because of the Earth's revolution around the Sun, the stars rise and set four minutes earlier every night. Four minutes each day adds up to about 30 minutes each week and two hours each month.

In six months, the stars that rise above your horizon at 8 PM tonight will rise 12 hours earlier, at 8 AM. And unless you live near the North or South Poles, the Sun has already risen by 8 AM, so you will not be able to see the stars.

This is why any stars that rise and set go through periods when they are not visible to us on Earth.

## 13.9 Star Wheels

"I understand now why star charts have to be labelled with a certain time of night," Mike said. "But why are they also labelled with a particular month?"

"Because the stars we can see at 9 PM in the summer are different from the stars we can see at 9 PM in the winter," Dad explained. "This is because the Earth is **revolving** around the Sun as well as rotating on its axis. You can see this in Figure 13.20 in your textbook. The Earth's path around the Sun is called its **orbit**, and it takes exactly one year for the Earth to travel all the way around the Sun and back to its starting position. This is what a year is: the time it takes the Earth to complete its orbit around the Sun.

"Now, at any point on the Earth's orbit, we can only see the stars that are behind the Earth. This is because we can only see the stars at night, and it is only night when our part of the globe is facing away from the Sun. When we are facing towards the Sun, it is daytime, so we can't see any stars.

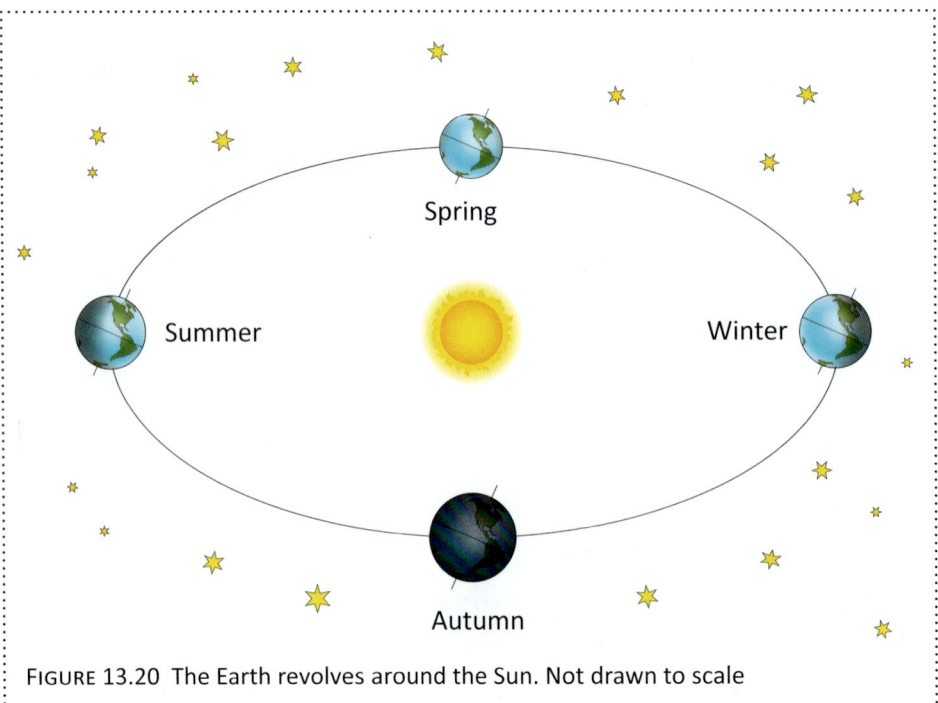

FIGURE 13.20 The Earth revolves around the Sun. Not drawn to scale

"As the year goes by and the Earth travels farther along its orbit around the Sun," Dad continued, "we start to see different stars at night. This is because we are moving around the Sun, so the stars that used to be behind the Sun are now behind the Earth. That is why we see different stars at different times of the year."

"I don't think I can keep track of all this moving around!" Mike groaned.

"Don't worry—keeping track of the motion of the stars is difficult for everyone else, too," Dad reassured Mike. "That's why stargazers often use a special type of star chart, called a star wheel."

Dad produced a star wheel from his stack of star-gazing resources and handed it to Mike. "As you can see, a star wheel has two parts: a disk that shows the positions of the stars and is labeled with the months of the year, and a frame that covers part of the chart and is labeled with the different hours of the day.

"To use a star wheel," Dad explained, "you turn the disk and the frame in opposite directions until the month in which you are star-gazing matches up with the time of night you want. For instance, if you are star-gazing at 10 PM in June, you should twist the two halves of the star wheel until '10 PM' on the frame matches up with 'June' on the disk."

"And then the window in the frame shows you what the sky will look like at that time of night in June?" Mike asked.

"Exactly," Dad answered. "The edge of the window represents the horizon, and its sides are labeled 'North,' 'South,' 'East,' and 'West.' To find a star that is near the northern horizon, you should face north and hold the star wheel so the northern horizon is at the bottom of the wheel. Then you can hold the wheel above your head and use it just like a regular star chart."

"Great!" Mike said. "Let's go try it!"

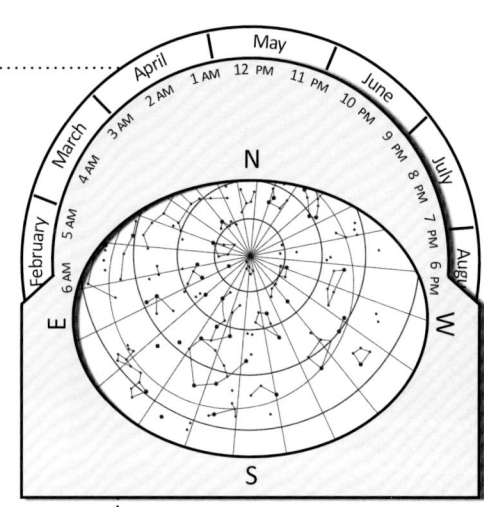

FIGURE 13.21 This star wheel is set to 10 PM in June.

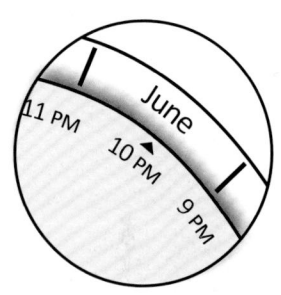

To **revolve** means to travel in a circle around something else. One revolution of the Earth around the Sun is the same as one year.

The Earth's **orbit** is its path around the Sun.

 Unit 3 Activity #3

 Workbook pgs. 75-76

## 13.10 Latitude Lines and the Height of Polaris

"As you can see by comparing Figures 13.16, 13.17, and 13.18 in your textbook," Dad said the next evening, "your view of the sky changes depending on where you are between the Equator and the North Pole. If you live close to the North Pole, Polaris is high in the sky; if you live close to the Equator, Polaris is low in the sky.

"Because Polaris is higher or lower in the sky depending on how far you are from the Equator, early explorers and sailors used Polaris to find their way at night. Depending on how high Polaris was in the sky, travellers knew whether they were north or south of where they wanted to go."

"We should have used Polaris to find our way during our virtual road trip!" Mike exclaimed.

"Hey, we never lost our way," Dad said, pretending to be hurt. "At least, not much."

"I wish we could really go on a road trip so I could learn how to find my way by the stars," Mike said.

"We aren't travellers, but we can still use Polaris to estimate how far we are from the Equator. Let's go outside and I'll show you how."

Out in the backyard, Dad continued. "To find out how far you are from the Equator, all you need to do is measure the distance between Polaris and the northern horizon, using your fists and little fingers."

"But I can't see the horizon. There are too many trees and hills in the way," Mike objected.

"That's all right," Dad said. "Just estimate where the horizon would be if we lived on flat ground."

"Okay," Mike said. "Polaris is four fists and three finger-widths above the horizon."

"Great!" Dad said. "As you've learned in math, scientists and mathematicians often divide circles into three hundred and sixty degrees, which we write as 360°. The night sky looks like a great, round dome, so we can divide it up into 180°, which is half a circle. Since the distance from the horizon to the highest point in the sky is half of the dome, scientists would say that it is a distance of 90°."

"Okay," Mike said, hesitantly. "What does this have to do with my fists and fingers?"

"When you hold your fist against the sky," Dad explained, "your fist covers approximately 10° of the sky, and the width of your little finger covers about 1°. So can you tell me how many degrees Polaris is from the horizon?"

"If one fist equals 10° and one finger-width equals 1°, then four fists and three finger-widths is 43°," Mike calculated. "So Polaris is 43° above the horizon?"

FIGURE 13.22 The distance from the horizon to the highest point of the sky is 90°.

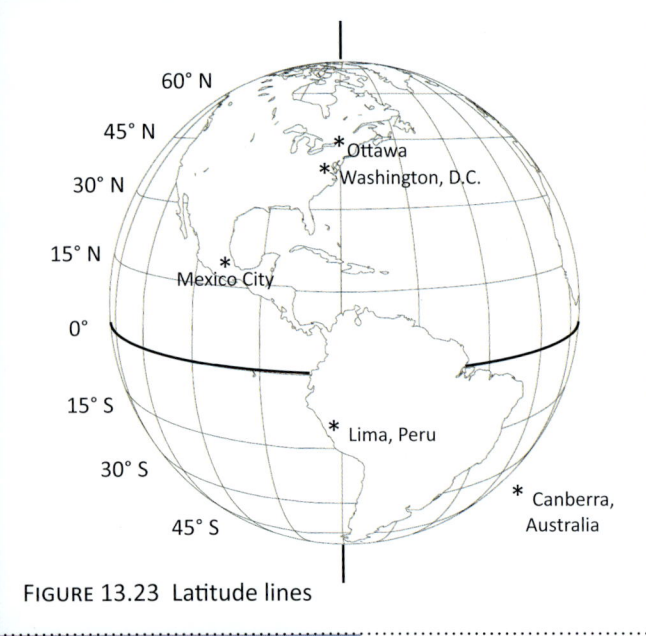

FIGURE 13.23 Latitude lines

**Latitude** is the way that scientists measure a location's distance from the Equator. Each degree of latitude is about 69 miles farther from the Equator.

If you already know your latitude, you can use it to confirm that a certain star is Polaris by measuring the distance between the star and the horizon in fists and finger-widths.

"Perfect!" Dad said.

"But how does that tell us where we are on the globe?" Mike asked.

"Do you remember learning about latitude lines in Chapter 7?" Dad asked. "**Latitude** is the way that scientists measure a location's distance from the Equator. Latitude line 0° is the Equator, and the North and South Poles are 90° north and south of the Equator."

"So everywhere in between the Equator and the Poles is at a latitude line from 1° to 89°?" Mike asked.

"Exactly," Dad said. "For instance, Mexico City is located at 19° N, Washington D.C. is located at 38° N, and Ottawa is located at 45° N. Lima, Peru is located 12° below the Equator, so its latitude line is 12° S. The latitude line of Canberra, Australia is 35° S, which means it is located even farther below the Equator.

"And here is the exciting part," Dad continued. "Your latitude is always the same as the height of Polaris in your sky. So since you measured Polaris to be 43° above the horizon, this tells us that we live at 43° N latitude!" Dad exclaimed.

"But why would I want to know that I live at 43° N latitude?" Mike asked.

"Look at it this way," Dad said. "If you were a ship's captain in the 17th century sailing southwest from England to America, how would you know when you had sailed far enough south? You would know because previous travellers could tell you that the colony for which you were heading was located at 39° N. At that latitude, Polaris would be almost four fists above the northern horizon, and you would know that you should stop sailing south and head west for land."

"Okay, I agree. Latitude is pretty awesome," Mike admitted.

## 13.11 The Celestial Clock

Polaris and the Big Dipper can be used to tell time at night. The Big Dipper makes a complete circle around Polaris once every 24 hours, so we can imagine Merak and Dubhe, the pointer stars, as the hour hand of a clock.

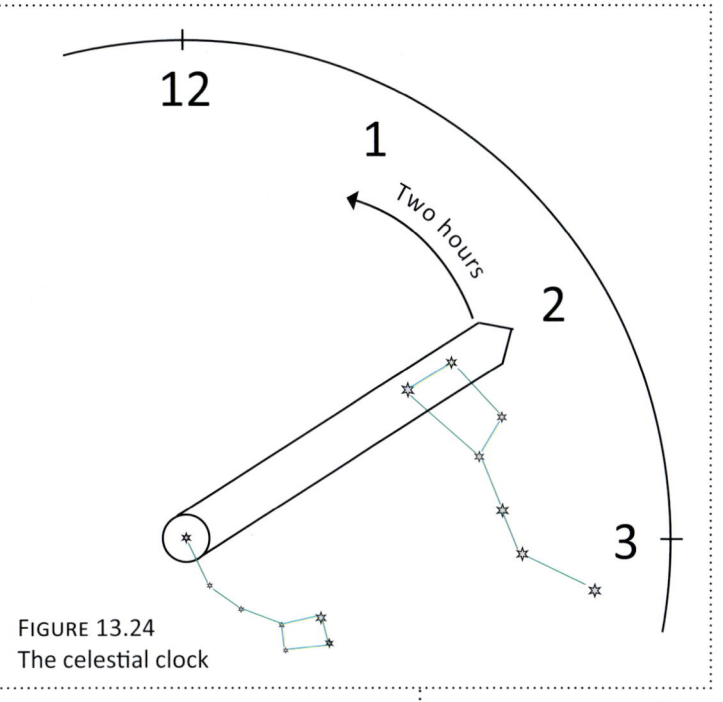

FIGURE 13.24
The celestial clock

There are two main things to remember when using Merak and Dubhe to calculate the time of night. First, Merak and Dubhe move counterclockwise around Polaris, not clockwise like the hands of a normal clock. Second, the **celestial** clock is a 24-hour clock, not a 12-hour clock. On a 12-hour clock, it takes one hour for the hour hand to move from the 2 o'clock position to the 3 o'clock position, but it takes Merak and Dubhe two hours to move the same distance. Of course, since the celestial clock moves counterclockwise, Merak and Dubhe move from the 3 o'clock position to the 2 o'clock position.

Once you get the hang of it, you can use the celestial clock at night instead of a watch to estimate how much time has passed since you started star-gazing. For instance, let's say Merak and Dubhe were in the 2 o'clock position[3] when you first came outside. This does *not* mean that you started star-gazing at 2 AM or at 2 PM. But it does mean that, if Merak and Dubhe are in the 1 o'clock position the next time you look, you can know that you have been star-gazing for two hours. This is because you know that it takes Merak and Dubhe two hours to travel one-twelfth of the distance around a circle.

**Celestial** means "having to do with the sky or the heavens."

Unit 3 Activity #4

Workbook pgs. 77-78

Star-Gazing Test pg. 78

---

[3] That is, the position in which the hour hand of a 12-hour clock would be at 2 o'clock. See Figure 13.24.

The International Space Station orbiting the Earth

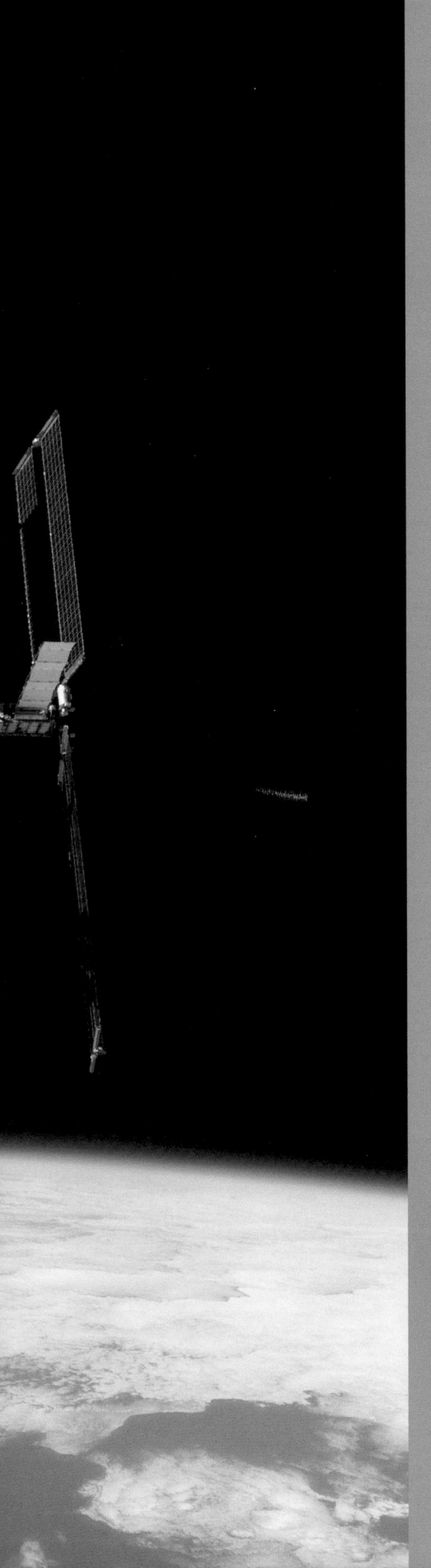

# 14
## Chapter

# Gravity, Orbits, and the Moon

Gravity explains the motions of the planets, but it cannot explain who sets the planets in motion.

— Sir Isaac Newton

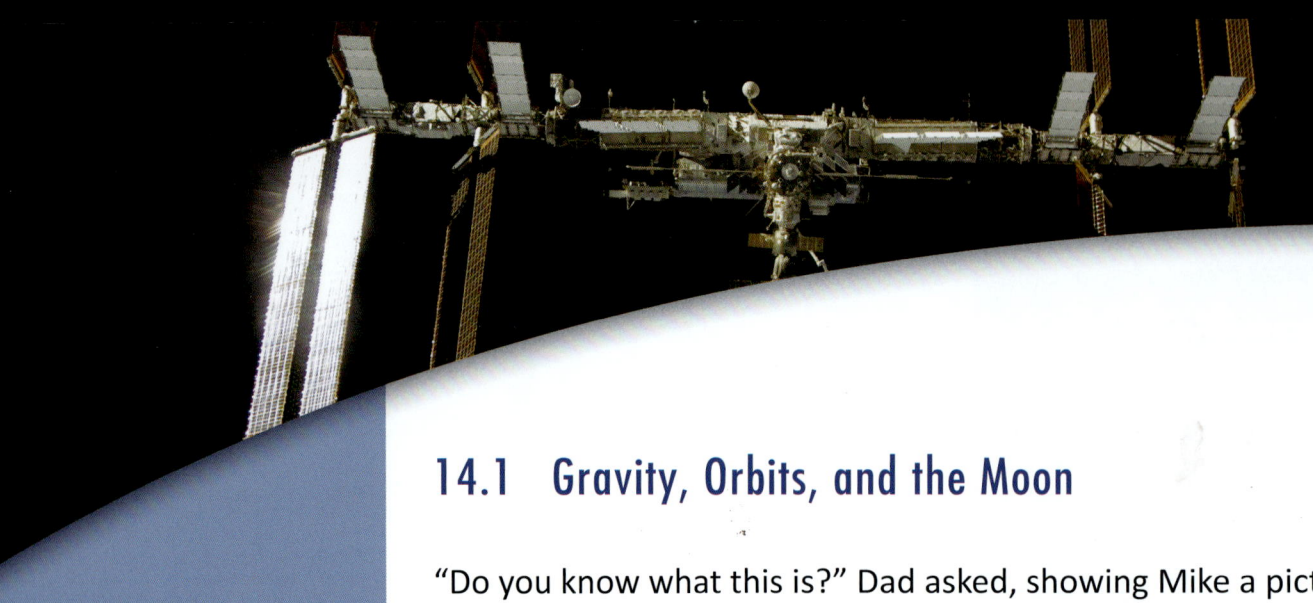

## 14.1 Gravity, Orbits, and the Moon

"Do you know what this is?" Dad asked, showing Mike a picture of the International Space Station **orbiting** the Earth.

"Is it some sort of spaceship?" Mike answered.

"It's the International Space Station, which is essentially a huge science **laboratory** in outer space," Dad said. "Astronauts perform all sorts of science experiments on board the International Space Station."

"It sounds like a great place to learn about outer space!" Mike said.

"And that's why the International Space Station, or the ISS as it is called, will be our virtual home base for our study of astronomy!" Dad announced.

"So to begin our study of outer space, imagine that we are strapped into a space craft, waiting for lift-off. Countdown is nearing completion: *T minus 5 seconds, T minus 4 seconds, minus 3, minus 2, minus 1, lift-off!* The space craft is thrust into the sky by its powerful rockets. All we can hear is a tremendous roaring

To **orbit** (verb) means to travel around a star or planet because of the gravitational pull of the star or planet.

An **orbit** (noun) is the path of an object around a star or planet.

A **laboratory** is a place where scientists perform experiments.

FIGURE 14.1
Space Shuttle Discovery Launch, 1988

as the space craft carries us away from the Earth at a terrific speed.

"After about 10 minutes, the space craft has reached its **orbit**, which is its path around the Earth. Its engine shuts off, and we 'coast' through space in perfect silence. Although the space craft is moving incredibly fast, it takes about two days for it to 'catch up' to the ISS, because the ISS is also orbiting the Earth at high speed.

"Meanwhile, we unbuckle our seatbelts and look out the window. We are far above the clouds, and we see entire countries and continents passing beneath us. From our height of 240 miles above the Earth, we can see the Earth as the round globe that it is. Instead of being flat, the horizon is curved. We can see the atmosphere as a soft, glowing layer above the horizon.

"Although we have an incredible view of the Earth, the strange feeling of weightlessness that we experience is even more astounding. When we try to walk, the push of our feet against the floor sends us flying towards the ceiling. I take off my launch helmet and set it on a chair, but it simply floats

FIGURE 14.2 Looking through the window of the ISS

Although we could not see the curvature of the Earth from space until the second half of the 20th century, human beings have long known that the Earth is round. In 240 B.C., a Greek astronomer calculated the circumference of the Earth with incredible accuracy. In the middle ages, artists often portrayed the Child Jesus as holding the *orbis terrarum*, the "globe of the world," in his hand. At the beginning of the *Summa Theologiae*, St. Thomas Aquinas comments that "both the astronomer and the physicist can prove that the Earth is round." Finally, in 1519, Ferdinand Magellan led the expedition that successfully sailed around the globe for the first time in history.

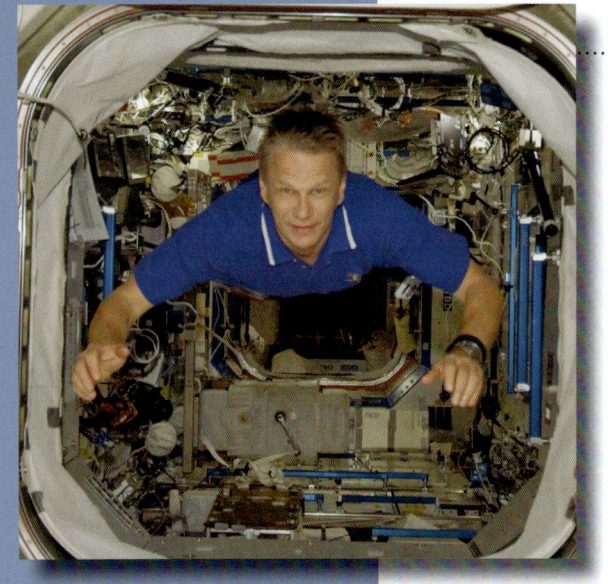

away. You stride forward to catch it for me, but the motion sends you somersaulting through the air. Strangely enough, you discover that it doesn't feel any different to be upside-down in space—whichever direction you are facing, you experience the same weightlessness. Eventually we learn how to 'swim' through the air, drifting around the room and guiding ourselves gently by the handles on the walls, floor, and ceiling.

"After about two days, our space craft docks at the space station," Dad continued. "Just as on the space craft, weightlessness affects everything we do. For instance, we always wear socks or soft shoes while we move about the cabin, because we don't want to damage any equipment as we fly through the air. When it is time to eat, we don't use utensils the way we do at home—the food would simply float off our forks! Instead we use spoons or chopsticks to guide the floating food into our mouths. At mealtimes, nuts and M&M's go flying through the air, and spilled fruit juice balls up and floats around the room! In fact, even plates and cups would float away on the ISS, so most food comes in packages that can be strapped or clipped to the table.

"Weightlessness even affects how we sleep at night. When it is time to sleep, we don't lie on a bed. Instead, we crawl into a sleeping bag that keeps us from floating away in our sleep. It's just as comfortable to sleep vertically as horizontally, so our sleeping bags are hung to the wall or the ceiling."

Mike had been listening eagerly to Dad's description of outer space. Here he posed a question: "Why do we feel weightless on the ISS? Is it because there isn't any gravity in outer space?"

"On the contrary," Dad said, "gravity is definitely present in outer space. Gravity is what keeps the Earth in its orbit around the Sun and the Moon in its orbit around the Earth. In fact, gravity is what keeps the space craft and the space station in their orbits around the Earth."

"Then why do we feel weightless in outer space?" Mike asked, puzzled.

"That is a good question with a complicated answer. To understand the answer, we first need to understand what we mean when we say that something is in orbit.

"If you throw a ball as hard as you can, it will follow the curve of the Earth and eventually fall to the ground. (Figure 14.3)

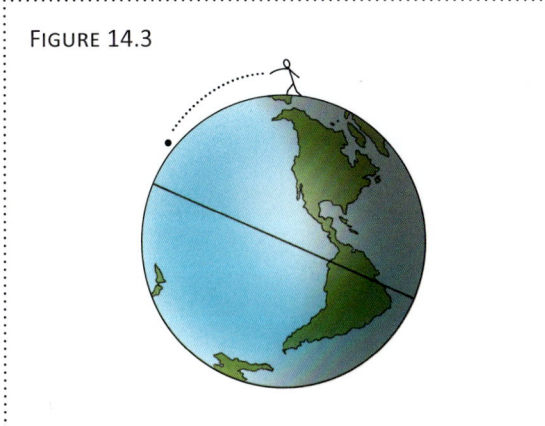

FIGURE 14.3

"If a giant threw the same ball, he could throw it a lot farther, and the ball would stay in the air longer before falling to the ground, still following the curve of the Earth. (Figure 14.4)

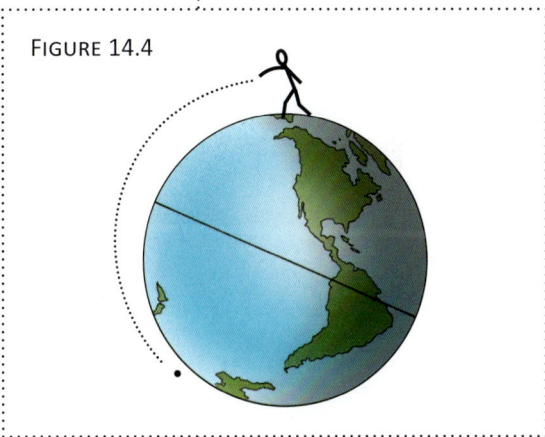

FIGURE 14.4

"And if you found the giant's father, he might be able to throw the ball so hard that it would end up travelling all the way around the Earth so he could catch it on the other side. (Figure 14.5)

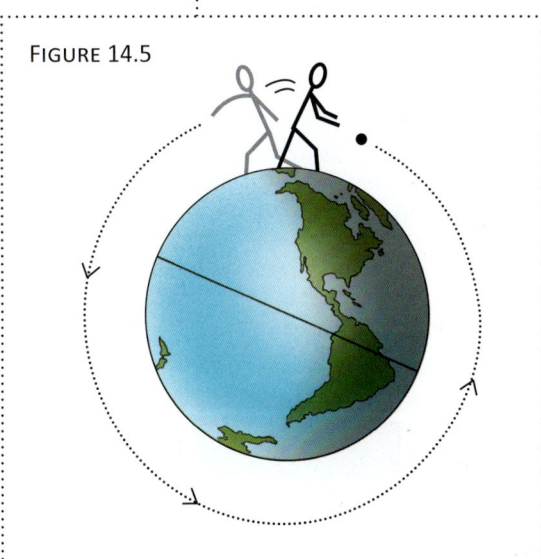

FIGURE 14.5

"That's what it means for an object to be orbiting the Earth. When a spacecraft is sent into orbit, it is shot into space so hard that it just falls all the way around the Earth—over and over again, as the Earth curves away from it. The curve of the spacecraft's path is the same as the curve of the Earth's surface, so the spacecraft never hits the ground.

"In other words, a spacecraft in orbit is in a state of continuous freefall. This is what makes astronauts feel weightless—they are falling and falling and falling all the way around the Earth."

"Falling gives you a feeling of weightlessness?" Mike asked.

"Yes," Dad affirmed. "You experience the beginning of weightlessness and freefall when an elevator begins to descend. If the cable were suddenly to break and the elevator began to plummet to the ground, you would experience the same freefall as the astronauts do on the ISS. You and the elevator would both be falling at the same speed, so you would appear to be floating in mid-air. If the elevator kept falling for an hour or two, you would have time to turn somersaults in the air, take a nap upside down, send objects floating through the air with a touch of the finger, and do all the other things that people often imagine can only be done in outer space."

"It wouldn't be much fun when the elevator hit bottom, though," Mike commented.

"Which is why people don't line up outside broken elevators to go into freefall," Dad chuckled. "As far as I know, outer space is the only place where you can experience freefall and weightlessness for an extended period of time."

FIGURE 14.6 If the cable on an elevator were to break, you would experience the same freefall and weightlessness as astronauts do in outer space.

Our **solar system** is a collection of planets, asteroids, comets, and other objects orbiting the Sun.

## 14.2 Gravity in the Solar System

Every mass exerts a "pull" on every other mass because of the force called gravity. Since your body possesses mass, it exerts a pull on the Earth in the same way that the Earth exerts a pull on your body. But the gravitational pull of your body is too small to notice because you have so little mass compared to the Earth.

The more massive an object is, the stronger its gravitational pull is. The Earth exerts a strong pull because it is very massive. The strong gravitational pull of the Earth is the reason that balls, paper airplanes, and children fall down instead of up. And as Mike learned on his virtual trip to the International Space Station, the gravitational pull of the Earth is also the reason that the ISS, space crafts, and the Moon travel in orbits around the Earth.

Our **solar system** is a collection of planets, asteroids, comets, and other objects orbiting the Sun. Since the Sun is nearly one hundred times more massive than everything else in the solar system, it has a stronger gravitational pull than the rest of the solar system put together. This is why everything in the solar system travels around the Sun. Most objects in the solar system, such as planets, asteroids, and comets, travel in direct orbits around the Sun. The various moons in the solar system do not orbit the Sun directly, but they travel around the Sun by accompanying the planets in *their* orbit around the Sun.

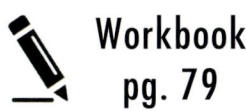

Workbook pg. 79

### OUTER SPACE

Living in outer space is quite different from living on Earth. For one thing, there is no air in outer space. When astronauts go "space walking" outside their spaceship, they have to wear a space suit and oxygen tank.

The lack of air makes shadows very cold and sunlight very hot in space. On Earth, sunshine warms the air as well as the side of your face, the ground, and the sunny side of boulders. The warm air helps spread the heat to places that are in shadow. Since this cannot happen without air, shadows are very cold in outer space. Without an atmosphere, the Earth would also have extremely hot temperatures in the sun and extremely cold temperatures in the shade.

Perhaps you have seen science fiction movies in which planets or spaceships explode with thunderous sound effects. If those explosions actually took place in outer space, they would be completely silent because sound cannot travel through empty space. Without air to carry the sound waves to our ears, we could not hear even the biggest explosions.

FIGURE 14.7 Buzz Aldrin on the Moon

## 14.3 Planet Earth's Moon

"Dad, I wish planets didn't have moons," Mike said. "The solar system would be so much easier to understand if everything just orbited the Sun."

"I know what you mean," Dad admitted. "If it weren't for the Moon, all you'd have to keep track of would be the Earth's rotation on its axis and its revolution around the Sun. The Moon makes things a lot more complicated.

"The Moon is quite important for life on Earth, though," Dad said. "When we studied biomes you learned that the Earth's axis is tilted to one side. The tilt of the Earth's axis is what causes the seasons, so you can understand why it is so important for the tilt to remain the same."

"I can?" Mike said.

"Yes," Dad assured him. "The Earth's axis is tilted by 23.5°, which is just the right tilt to give year-round warmth to the tropics and four seasons to the temperate zones. Imagine what would happen to the plants and animals in the tropics if the tilt of the Earth's

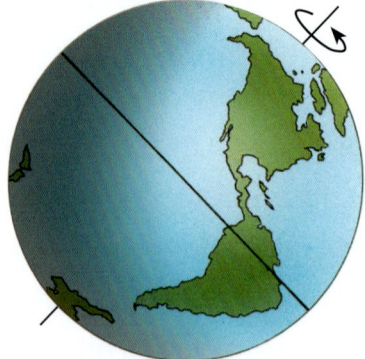

The Earth with a 23.5° tilt    The Earth with a 45° tilt

FIGURE 14.8 Since the tilt of the Earth's axis causes the seasons, a more extreme tilt would result in more extreme seasons.

axis suddenly changed to 45°. The tropics would no longer receive direct sunlight all year round, so they would have colder winters and warmer summers. Since tropical plants and animals are not equipped for this type of weather, many of them would die.

"A more extreme tilt in the Earth's axis would also affect the rest of the globe. Every climate zone would start to have warmer summers and colder winters. Imagine how hard it would be for polar bears and arctic foxes to survive temperatures of 70 or 80°F with their thick fur coats! Parts of the globe that are now in the temperate zone, such as most of the U.S., would endure rainforest summers and boreal forest winters."

FIGURE 14.9 The small white speck in the upper left corner is the ISS. Even though it looks like the ISS is approaching the Moon to land on it, it is actually very far away.

"I didn't realize how carefully God fine-tuned our seasons," Mike said. "But what does the Moon have to do with the tilt of the Earth's axis?"

"The Earth has a very large moon compared with most planets of its size," Dad explained. "Because the Moon is so large, its gravitational pull is strong enough to stabilize the tilt of the Earth's axis. Without the consistent gravity of the Moon, the Earth would 'wobble' on its axis, and we would have a different type of climate every thousand years. As it is, the Earth's seasons have remained the same for thousands of years—thanks to the Moon."

"I'm flabbergasted!" Mike exclaimed. (Mike had run across the word "flabbergasted" a few weeks ago in a library book, and it had quickly become one of his favorite terms.) "I didn't know how important the Moon was. Astronomy would be simpler without the Moon, but life would be much more difficult!"

"And less awe-inspiring," Dad added. "Besides making sure we have consistent seasons, the Moon is also a beautiful part of the night sky. Plus, lunar and solar eclipses are only possible because of the Moon."

"Do we get to learn about eclipses?" Mike asked. "The Moon is much more interesting now that I know how important it is."

"I'm glad of that," Dad said, "because one of the reasons we are learning about gravity and orbits is so that we can understand eclipses, the phases of the Moon, and the tides. We'll start by learning about the phases of the Moon."

## 14.4 Phases of the Moon

The **phases of the Moon** are the different ways we see the Moon throughout the month.

The fact that the Moon is orbiting the Earth explains why the Moon goes through different phases. The **phases of the Moon** are the different ways we see the Moon throughout the month. A *full moon* is the phase in which we see the Moon as a full circle, and a *new moon* is the phase in which we cannot see the Moon at all. A *crescent moon* is the phase in which we see only a sliver of the Moon, a *quarter moon* is the phase in which we see half of the Moon, and a *gibbous moon* is the phase in which we see three-quarters of the Moon.

As the Moon orbits the Earth, the side of the Moon that faces the Sun is always fully illuminated and the other half is always in darkness (Figure 14.11). What changes during the month is how much of the Moon's bright side is visible from Earth.

FIGURE 14.10 The phases of the Moon

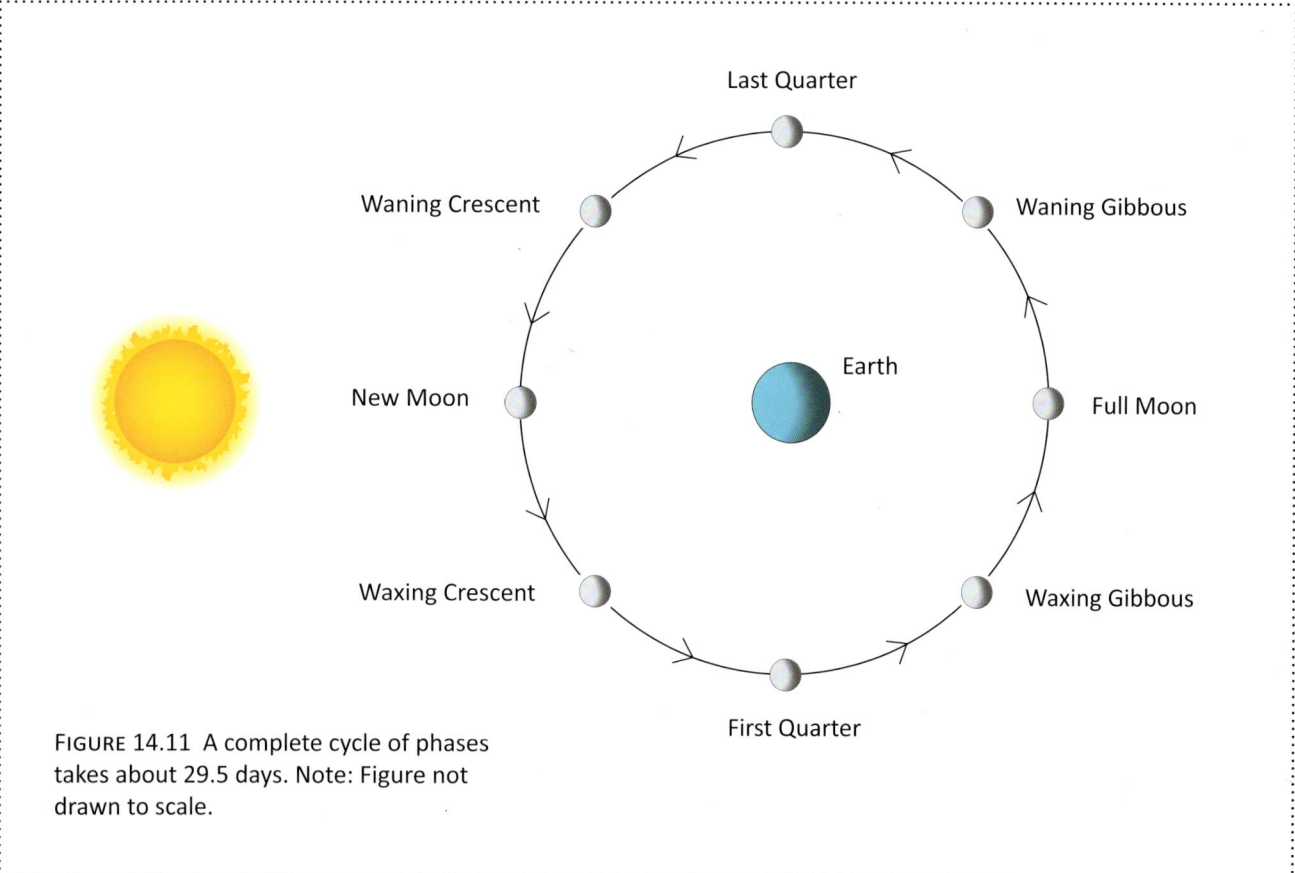

FIGURE 14.11 A complete cycle of phases takes about 29.5 days. Note: Figure not drawn to scale.

A full moon occurs when we can see the entire bright side of the Moon. As you can see in Figure 14.11, a full moon occurs when the Moon is on the far side of the Earth. The Moon does not remain full for more than a few days. As the Moon slowly orbits the Earth, the bright side of the Moon begins to be turned away from the Earth, and we see a **waning** gibbous moon.

When the Moon has waned until only half of it is visible from the Earth, we say that the Moon is a last quarter moon. After the last quarter, the Moon's bright side turns away from the Earth even more, until all we can see is a waning crescent moon.

Eventually, the thin sliver of the waning crescent disappears from view and we cannot see the Moon at all. This phase is called a new moon, and it occurs when the bright side of the Moon is turned away from the Earth entirely. During a new moon, the side of the Moon facing the Earth is completely dark.

The term **waning** (WAYN-ing) is used to refer to the Moon during the period when we see less and less of it each day.

As the Moon continues in its orbit, the new moon **waxes** back into a crescent moon. After a few days, the waxing crescent moon becomes a first quarter moon, and the first quarter moon eventually becomes a waxing gibbous moon. After nearly a month, the Moon returns to its original position on the far side of the Earth. It is now a full moon once more!

To understand why the Moon waxes and wanes, find a ball and stand outside on a sunny morning. Holding the ball at arm's length, turn around slowly, keeping your eyes on the ball. How much of the bright side of the ball can you see when the Sun is behind you? How much of the bright side of the ball can you see when the Sun is in front of you? How much of the bright side of the ball can you see when the Sun is at your side?

The term **waxing** is used to refer to the Moon during the period when we see more and more of it each day.

Unit 3 Activity #5

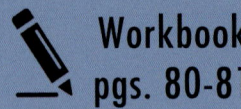
Workbook pgs. 80-81

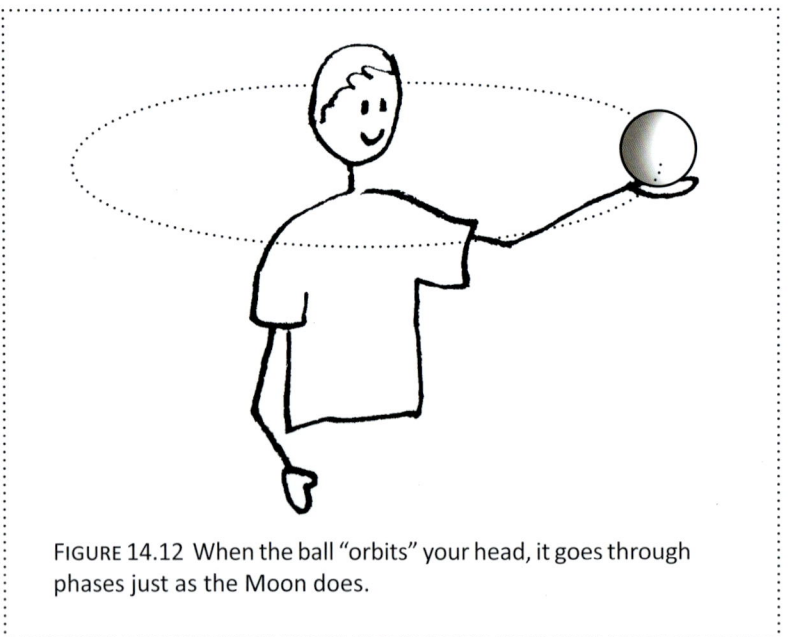
FIGURE 14.12 When the ball "orbits" your head, it goes through phases just as the Moon does.

## 14.5 Lunar Eclipses

The word "lunar" comes from the Latin word "luna," which means "moon." A **lunar eclipse** occurs when the Earth's shadow falls upon the Moon, throwing all or part of the Moon into darkness.

Lunar eclipses can only occur during a full moon because that is the only time the Moon passes directly behind the Earth (Figure 14.11). Lunar eclipses do not occur at *every* full moon, because the Moon's orbit is slightly tilted in comparison with the Earth's orbit, as you can see in Figure 14.13.

Since the Moon's orbit is tilted, the Moon is usually positioned slightly above or below the Earth, instead of passing through the Earth's shadow at every full moon. But several times a year, the Earth, Moon, and Sun are perfectly aligned, and then everyone on the dark side of the Earth gets to enjoy a lunar eclipse.

A **lunar eclipse** (LOO-ner ee-KLIPS) occurs whenever the Earth's shadow falls upon the Moon, throwing all or part of the Moon into darkness.

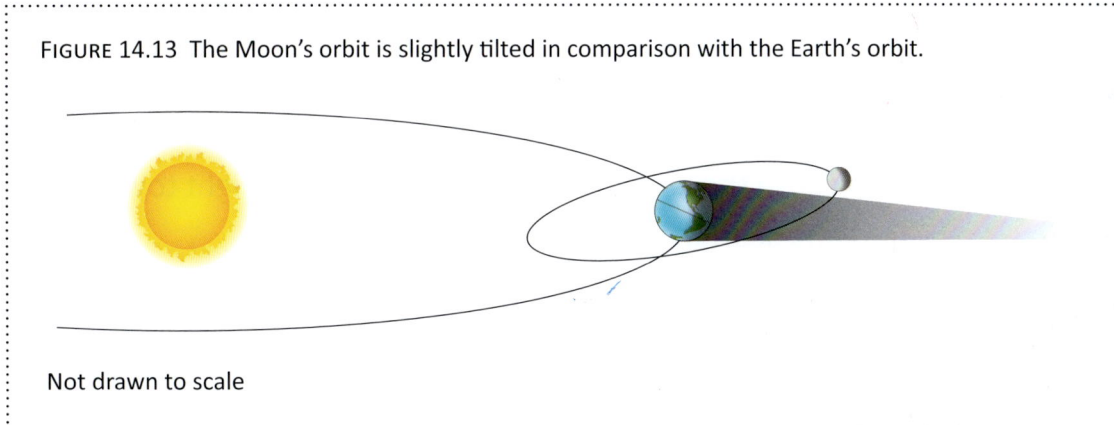

FIGURE 14.13 The Moon's orbit is slightly tilted in comparison with the Earth's orbit.

Not drawn to scale

**There are three kinds of lunar eclipses**: total lunar eclipses, penumbral lunar eclipses, and partial lunar eclipses. The reason there are different types of lunar eclipses is that the Earth's shadow has two parts, the penumbra and the umbra.

The **penumbra** (pin-UHM-bruh) is the large, outer region of a shadow in which the Sun's light is partially blocked.

The **umbra** (UHM-bruh) is the central, darker region of a shadow in which the Sun's light is totally blocked.

As you can see in Figure 14.14, the **penumbra** is a large, outer region in which the Sun's light is partially blocked. The **umbra** is a smaller, central region in which the Sun's light is totally blocked. How the Moon looks during a lunar eclipse depends on what part of the Earth's shadow it is passing through.

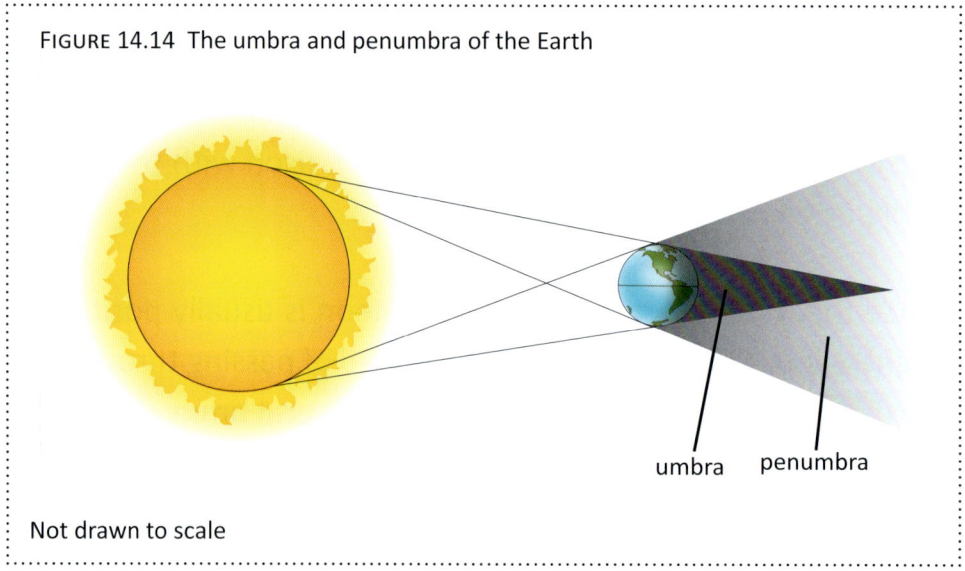

FIGURE 14.14 The umbra and penumbra of the Earth

Not drawn to scale

A **total lunar eclipse** occurs when the Moon passes through the Earth's umbra. During a total lunar eclipse, the entire Moon becomes dark for about an hour. The Moon takes on an eerie red hue because the Earth's atmosphere bends red light towards the Moon and scatters the other colors away.

Total lunar eclipse

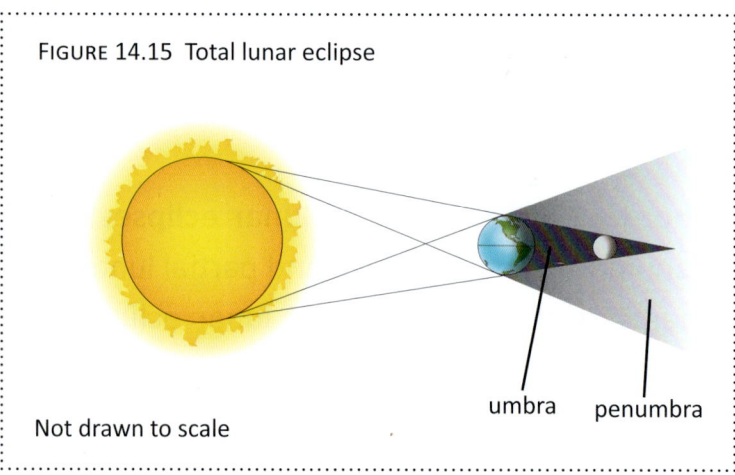

FIGURE 14.15 Total lunar eclipse

Not drawn to scale

A **penumbral lunar eclipse** occurs when the Moon passes through the Earth's penumbra instead of through its umbra. Since the Earth's penumbra only blocks part of the Sun's light, a penumbral lunar eclipse is not as impressive as a total or partial lunar eclipse. Instead of becoming totally dark, the Moon is darkened only slightly.

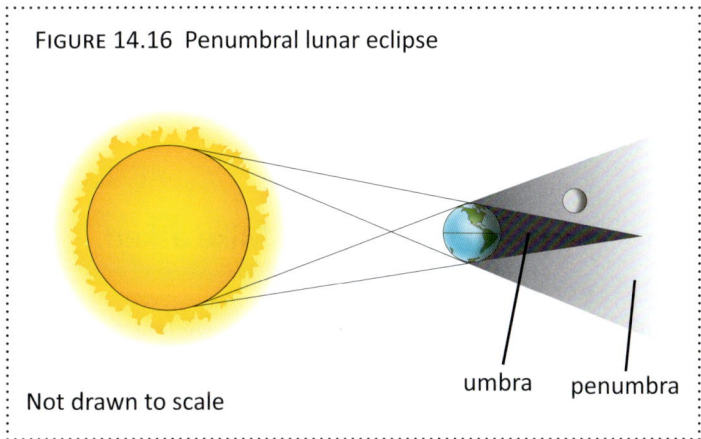

FIGURE 14.16 Penumbral lunar eclipse

Penumbral lunar eclipse

A **partial lunar eclipse** occurs when part of the Moon passes through the Earth's umbra and the rest remains in the Earth's penumbra. Since the Earth's umbra blocks all the light from the Sun, the part of the Moon within the Earth's umbra becomes very dark. The half of the Moon within the Earth's penumbra looks quite bright in comparison, because the penumbra only blocks a little of the Sun's light.

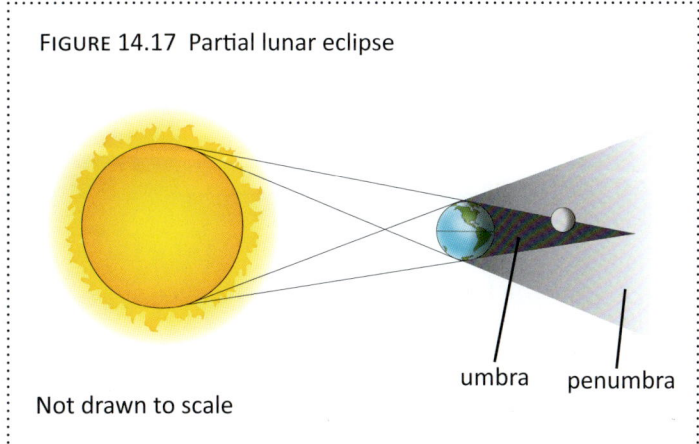

FIGURE 14.17 Partial lunar eclipse

Partial lunar eclipse

## 14.6 Solar Eclipses

"A solar eclipse is when the Moon blots out the Sun, right?" Mike asked.

"That's one way to put it," Dad said. "A **solar eclipse** occurs when the Moon passes in front of the Sun. Sometimes the Moon only covers up part of the Sun, but during a total solar eclipse, the Moon 'blots out' the entire Sun for about three minutes. This is called the **period of totality**."

"What is it like during the period of totality?" Mike asked.

"The Sun is completely hidden behind the Moon during this time," Dad explained, "and the sky becomes so dark that planets and brighter stars are visible in the sky. Birds stop singing and fly to their evening roosts, and frogs and crickets begin to croak and chirp.

A **solar eclipse** (SOL-er ee-KLIPS) occurs when the Moon passes in front of the Sun, casting a shadow on the Earth.

The **period of totality** is when the Sun is completely hidden behind the Moon during a total solar eclipse.

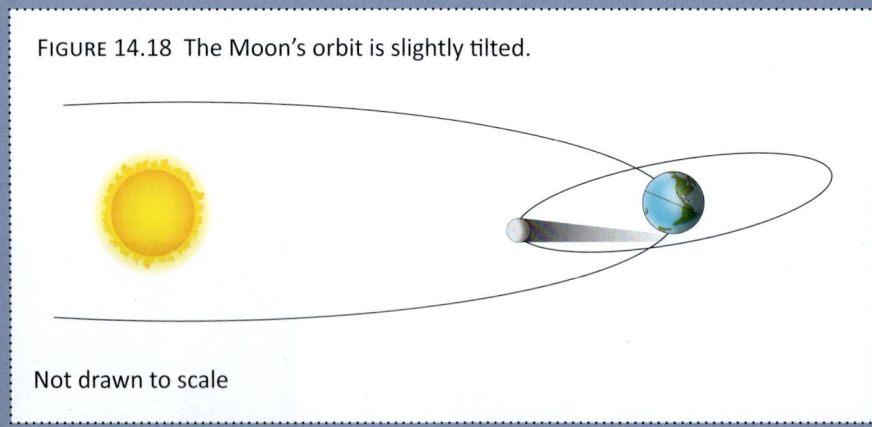

FIGURE 14.18 The Moon's orbit is slightly tilted.

Not drawn to scale

Solar eclipses can only occur during a new moon, because that is the only time the Moon passes directly between the Earth and the Sun (Figure 14.11). Solar eclipses do not occur at *every* new moon because the Moon's orbit is slightly tilted, so the Moon is usually slightly above or below the Earth. As a result, the Moon's shadow passes above or below the Earth and we do not see a solar eclipse.

"Most spectacular of all, a radiant halo called the **corona** appears around the Sun. The corona is a layer of extremely hot gases that surrounds the Sun. Except during a total solar eclipse, the corona cannot be seen without special equipment. This is because the fainter light of the corona is usually drowned out by the brightness of the central disk of the Sun."

"How often do total solar eclipses happen?" Mike asked eagerly. "Do you think we'll see one this year?"

"Total solar eclipses occur about once a year. The problem is that solar eclipses are only visible to certain parts of the Earth. This means that sometimes we have to wait many years for a total solar eclipse to be visible in the U.S."

"Everyone on the nighttime side of the globe can see a lunar

Total solar eclipse

The **corona** (kuh-RO-nuh) is a radiant halo of extremely hot gases that surrounds the Sun.

## Harvest Moon

The Harvest Moon is the full moon that occurs in September, during the harvest season. Farmers were grateful for the occurrence of a full moon during the harvest time because it provided them with enough light to continue harvesting their crops even after sunset. The Harvest Moon is not the only full moon that has a special name. Each month's full moon has its own name, usually derived from the activities traditionally associated with that time of year.

*January:* Wolf Moon
*February:* Snow Moon
*March:* Sap/Lenten/Worm Moon
*April:* Grass/Pink Moon
*May:* Planting Moon
*June:* Strawberry Moon
*July:* Thunder/Hay/Buck Moon
*August:* Grain/Sturgeon Moon
*September:* Harvest Moon
*October:* Hunter's Moon
*November:* Beaver Moon
*December:* Cold Moon

**Caution:** Never look at the Sun during an eclipse except during the few minutes of complete totality. The Sun is so bright that even gazing at a tiny sliver of the Sun can cause severe eye damage.

FIGURE 14.19 The Moon's umbral shadow as seen from the ISS during a solar eclipse

A **partial solar eclipse** occurs when the Moon's penumbra falls on the Earth.

eclipse," Mike said. "Why can't everyone on the daytime side of the globe see a solar eclipse?"

"Because a solar eclipse is caused by the Moon's shadow, and the Moon's shadow is much smaller than the Earth's shadow. In a lunar eclipse, the Earth's shadow is big enough to cover the entire Moon. But in a solar eclipse, the Moon's shadow is only large enough to cover part of the Earth.

"Here's a photo from the ISS that shows what a solar eclipse looks like from outer space. Everyone within the Moon's dark umbra will see a total eclipse."

"Oh no!" Mike exclaimed. "Don't tell me the Moon has an umbra and penumbra, too!"

"I know the names are confusing," Dad said, "but you can't understand the three types of lunar and solar eclipses without knowing about umbral and penumbral shadows."

"Are the three types of solar eclipses the same as the three types of lunar eclipses?" Mike asked.

"They are similar but not exactly the same," Dad said. "The three types of solar eclipses are total solar eclipses, partial solar eclipses, and annular solar eclipses."

"What's special about each one?" Mike asked.

"A **partial solar eclipse** occurs when the Moon's penumbra falls on the Earth. Anyone within the Moon's penumbral shadow will see a partial solar eclipse, because the penumbra is the region in which Sun's light is partially blocked by the Moon.

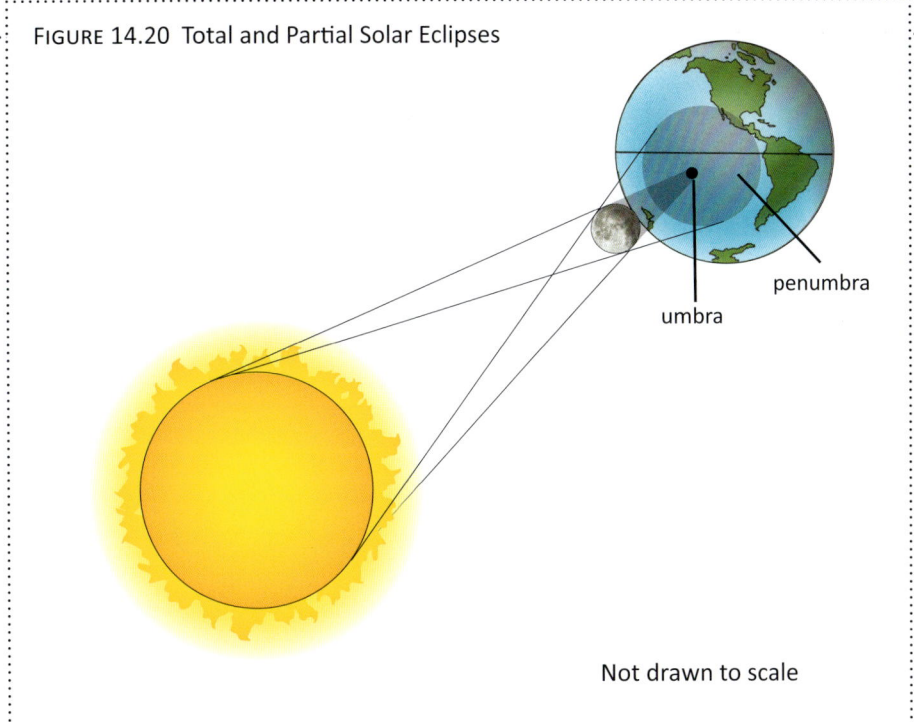

FIGURE 14.20 Total and Partial Solar Eclipses

umbra penumbra

Not drawn to scale

A **total solar eclipse** occurs when the Moon's umbral shadow falls on the Earth.

An **annular solar eclipse** occurs because the Moon's orbit around the Earth is not perfectly circular.

During a partial solar eclipse the Sun looks like an orange, glowing crescent moon.

"A total solar eclipse occurs when the Moon's umbral shadow falls on the Earth. People within the Moon's umbral shadow see a **total solar eclipse**, because the umbra is the region in which the Sun's light is completely blocked by the Moon. Of course, the Moon's umbra is surrounded by its penumbra, so while people in the Moon's umbra are seeing a total eclipse, people in the Moon's penumbra can see a partial eclipse.

"The third type of solar eclipse is an **annular solar eclipse**. Annular solar eclipses occur because the Moon's orbit around the Earth is not perfectly circular. The Moon's orbit is slightly oval, so its distance from the Earth varies from 221,000 to 252,000 miles," Dad said, glancing at his notes.

"When the Moon is near the Earth, at 221,000 miles away, it is just the right size to 'blot out' the Sun completely. This is when total and partial solar eclipses can occur. But

Partial solar eclipse

Total solar eclipse

Annular (AN-yuh-ler) solar eclipse

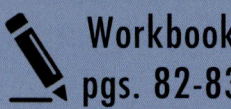

Workbook pgs. 82-83

when the Moon is far from the Earth, at 252,000 miles, it is too small to 'blot out' the entire Sun. When an eclipse occurs while the Moon is far from the Earth, the Moon blots out most of the Sun, but viewers can still see the Sun in a bright ring around the Moon. This is called an annular solar eclipse."

## 14.7  Gravity and the Tides

"What can you tell me about gravity?" Dad quizzed as he and Mike enjoyed their weightlessness aboard the ISS by doing virtual somersaults through the air.

"Gravity is a force that acts on matter," Mike said. "Every mass exerts a gravitational 'pull' on every other mass."

"Excellent," Dad praised. "Although scientists do not know exactly what gravity is, they can calculate exactly how strong gravity is between any two objects. To make this calculation, they only need to know two facts: the mass of the objects involved and the distance between the objects."

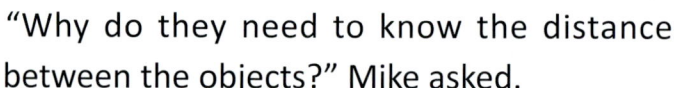

"Why do they need to know the distance between the objects?" Mike asked.

"As you know, the strength of the gravity between two objects changes depending on their mass. It also changes depending on how close they are to each other. For instance, the farther you travel from the surface of the Earth, the weaker its gravitational pull becomes."

"So is the Earth's gravity weaker at the top of

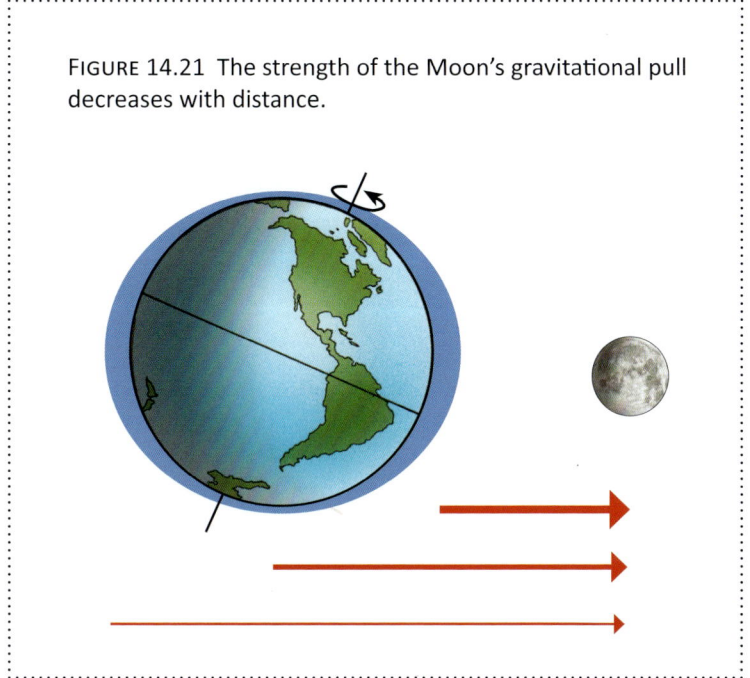

FIGURE 14.21 The strength of the Moon's gravitational pull decreases with distance.

*Not drawn to scale. The oceans and tidal bulges are much smaller in reality, and the Moon is much farther away.*

Mount Everest than it is at sea level?" Mike asked.

"Yes, but not by any noticeable amount," Dad answered. "The tides are the most noticeable way that we are affected by gravity decreasing over a distance."

"The tides?" Mike asked.

"Yes," Dad answered. "The tides are caused by the Moon's gravity 'pulling' on the oceans. Because of the Moon's gravitational pull, the water in the ocean collects in two 'bulges' on the Earth's surface. You can see this in Figure 14.21 in your textbook."

"I understand how that works," Mike said, confidently. "The Moon's gravity pulls on the ocean, so the ocean water forms a bulge on the side facing the Moon."

"That's partly right," Dad said. "But why are there two bulges in Figure 14.21, one on the side facing the Moon, and one on the side opposite the Moon?"

"Oh," Mike said. "I don't know. That doesn't seem to make sense."

"The reason is that the strength of the Moon's gravitational pull decreases with distance. The Moon pulls hardest on the side of the Earth that is facing it, and it pulls least hard on the side of the Earth that is opposite it (Figure 14.21)."

"How does this cause two bulges on the Earth?" Mike asked.

"The first thing to understand is that the Moon isn't only pulling on the Earth's oceans. The Moon is also pulling on the solid Earth itself. When gravity pulls on something that is very large, the object tends to become stretched, because the near side of the object is pulled harder than the far side of the object."

"Because the force of gravity is strongest when you are nearby and weakest when you are far away?" Mike asked.

"That's right," Dad said. "In the case of the Earth, the Moon pulls hardest on the near oceans, less hard on the solid Earth, and least hard on the far oceans. This is what causes the tides. The oceans on the near side of the Earth bulge because the Moon pulls harder on them than it does on the solid Earth. The oceans on the far side of the Earth bulge because they are not pulled towards the Moon with as much force as the rest of the Earth."

"Since there are bulges on each side of the Earth, are there high tides on both sides of the Earth at once?" Mike asked.

"That's right," Dad said. "High tide occurs whenever a particular part of the Earth is facing the Moon, *and* when that part of the

Earth is facing away from the Moon. And because the Earth is rotating on its axis, every place on Earth faces towards the Moon and away from the Moon once each day."

"And that must be why there are two high tides every day!" Mike concluded.

"Exactly," Dad said.

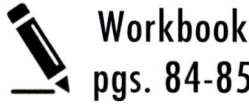

Workbook pgs. 84-85

The Cat's Eye Nebula, one of the first planetary nebulae discovered

# Chapter 15

# The Sun and the Other Stars

> Those who study the stars have God for a teacher.
> — Tycho Brahe

## 15.1 The Aurora Borealis from the ISS

"I'm ready for astronomy class," Mike said, joining Dad in the living room after finishing up the dinner dishes.

"Good!" Dad exclaimed. "I wouldn't want you to miss seeing the aurora borealis from outer space."

"The aurora borealis is the Northern Lights, right?" Mike asked. "I remember we saw a little one on the horizon a few years ago, but I wish I could see a really big **aurora**."

"Well, now you can," Dad said, "at least virtually. Astronauts onboard the International Space Station have an incredible view of

An **aurora** is a display of lights caused by the interaction of the solar wind with the Earth's magnetic field.

FIGURE 15.1 Aurora borealis photographed from the ISS

both the **aurora borealis** and the **aurora australis**—that is, the Northern and Southern Lights."

"The South Pole has an aurora, too?" Mike asked.

"Yes," Dad answered. "Aurorae occur at both the North and South Poles. The one in the north is called the aurora borealis, which means 'northern dawn,' and the one in the south is called the aurora australis, which means 'southern dawn.'"

"The aurora isn't really a dawn, though, right?" Mike asked.

"Absolutely not. The aurora borealis and the aurora australis are caused by the interaction of the solar wind with the Earth's magnetic field."

"The interaction of the *what* with the *what*?" Mike said.

"The interaction of the solar wind with the Earth's magnetic field," Dad repeated.

"Right now I have no clue what those are, but I suspect I'll know all about them after tonight," Mike grinned.

"Your suspicion is correct," Dad chuckled. "Our theme for tonight is the aurora borealis and its causes."

FIGURE 15.2 Aurora borealis photographed from the ISS

The plural of *aurora* is *aurorae* (uh-ROAR-ee).

**Aurora borealis** (uh-ROAR-uh bor-ee-AL-is)

**Aurora australis** (uh-ROAR-uh aw-STRAY-lis)

> Each second, approximately one million tons of particles escape from the Sun in the solar wind. Since the Sun is so huge, this "leak" of particles is barely noticeable. Less than one-thousandth of the Sun's mass has been lost due to the solar wind over the last five billion years.

> The **solar wind** is a stream of energetic particles that escape the pull of the Sun's gravity and shoot into outer space.

## 15.2 The Solar Wind, the Earth's Magnetic Field, and Aurorae

"Let's start by learning about the solar wind," Dad said. "I'm sure you have already guessed that the solar wind is not like the winds we have on Earth."

"No, because there isn't any air in outer space."

"That's right," Dad said. "Unlike ordinary wind, which is a current of fast-moving air, the **solar wind** is a stream of energetic particles that the Sun 'blows' outward in all directions. This solar wind flows through the entire solar system."

"How does the Sun 'blow' particles towards us?" Mike asked. "It can't do it the way we blow out the candles on a birthday cake!"

"You're right, it can't," Dad agreed. "The Sun doesn't actually *blow* the solar wind towards us. Rather, the solar wind consists of protons and electrons that escape the pull of the Sun's gravity and shoot into outer space."

"So the Sun is losing matter and getting smaller?" Mike asked, a little concerned.

"Technically, yes," Dad admitted. "But the Sun's gravity is so strong that relatively few particles are able to escape into space. It takes extremely high speeds to escape the pull of the Sun's gravity. The solar wind is the result of 'speedy,' energetic protons and electrons that burst away from the Sun's gravity into outer space."

"So how does the solar wind cause an aurora?" Mike asked.

"By interacting with the Earth's **magnetic field**. Do you remember learning about magnetic fields at the beginning of the year, in the chapter on electricity and magnetism? The Earth itself is a huge magnet, so it is surrounded by a giant magnetic field much like the magnetic field around a bar magnet." (Figure 15.3)

"Is the Earth's magnetic field the same as its gravitational pull?" Mike asked.

"Not at all," Dad said. "The pull of the Earth's magnetism is completely different from the pull of the Earth's gravity. The Earth has a gravitational pull because of its mass, but the Earth's magnetic field is produced by the motions of the **molten iron** in the Earth's outer core." (Figure 15.4)

"If the Earth is a giant magnet, why don't nails and iron skillets and other magnetic materials stick to the ground?" Mike asked. "If you brought them close to a regular magnet they would stick."

Dad laughed. "Luckily, the Earth's magnetic field is not strong enough to pull iron skillets out of our hands or make iron nails stick to the ground. This is because the magnetic part of the Earth, the molten outer core, is deep underground. Magnetic materials on the surface of the Earth are not completely unaffected by the Earth's magnetic field, though. For instance, in a magnetic compass, the magnetic compass arrow turns to point north because it is lining itself up with the Earth's magnetic field.

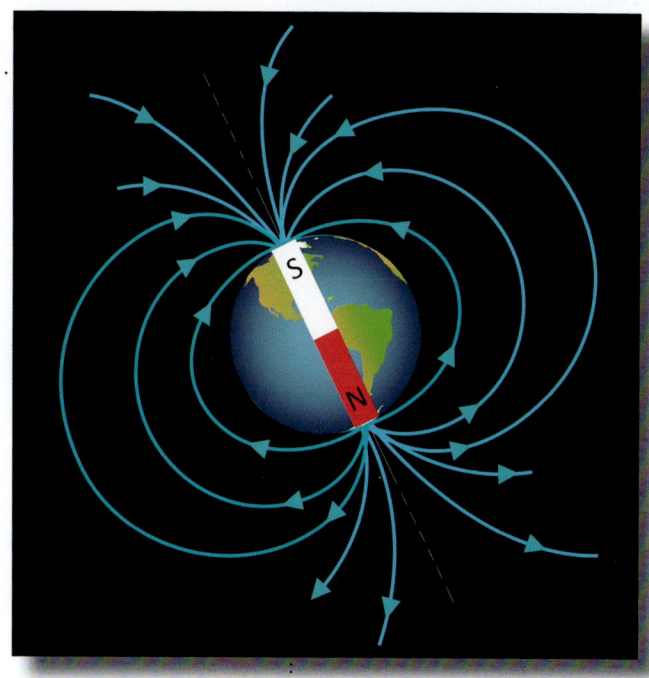

FIGURE 15.3 Earth's magnetic field

The Earth's **magnetic field** is produced by the motions of the molten iron in the Earth's outer core.

**Molten iron** is iron in its melted, or liquid, state.

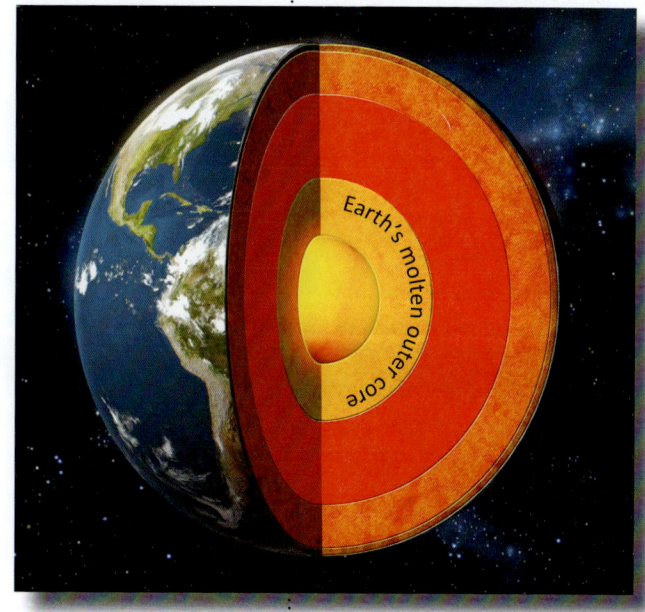

FIGURE 15.4

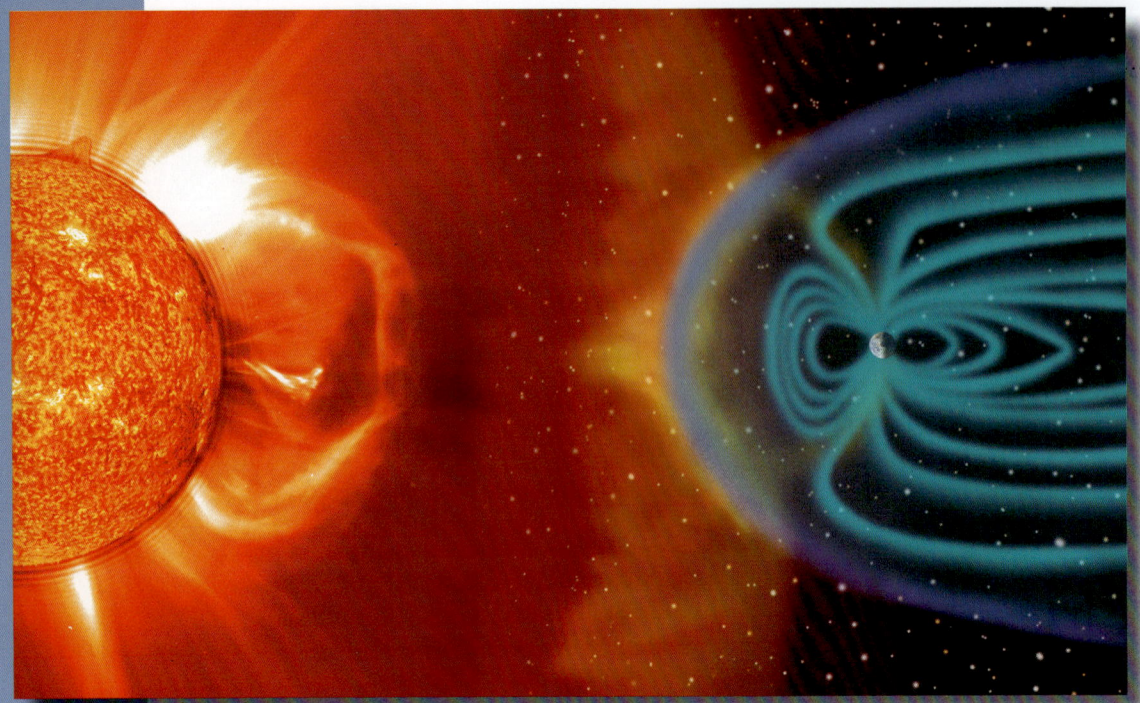

FIGURE 15.5 The Earth's magnetic field protects it from the solar wind.

"And even though the Earth's magnetic field isn't strong enough to pull iron skillets out of our hands, it is strong enough to protect the Earth from the solar wind. The magnetic field reaches far out into space, and forms a magnetic shield that prevents most of the particles in the solar wind from striking the Earth. If it weren't for the Earth's magnetic field, the solar wind would have stripped away the Earth's atmosphere a long time ago. And without an atmosphere, the Earth would be just another lifeless space rock."

"That is such good planning for God to give Earth a magnetic shield to protect its atmosphere!" Mike exclaimed.

"And such beautiful planning," Dad added. "Not only did God protect the Earth with a magnetic field; He even arranged it so the solar wind would interact with the magnetic field to produce beautiful aurorae."

"How do aurorae work?" Mike asked.

"Well, the magnetic field shields the Earth from most of the solar wind, but some particles manage to penetrate the field. These particles are channelled towards the North and South Poles, where the magnetic field lines loop through the Earth's core (Figure 15.5).

"When the particles enter the Earth's atmosphere, they collide with atoms of oxygen and nitrogen in the air. The energy produced by these collisions is released in the form of colored lights. An aurora is produced by billions of these collisions, because each flash of light lasts only a single instant."

"Why are some of the lights blue and some green?" Mike asked, looking at the photos in his textbook.

"Different types of collisions result in different colored lights. Oxygen atoms release green or red light, and nitrogen atoms release blue or red. Green lights are the most common, but each aurora is unique. Sometimes the lights shine quietly in broad loops and curtains, and at other times they move and shimmer in the sky as if they were dancing."

"Wow! Someday I'm going to take a road trip up north to see the aurora borealis in person!" Mike exclaimed.

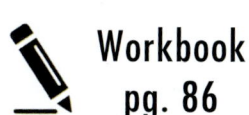

Workbook pg. 86

## 15.3 Solar Energy

The next evening, Dad asked, "Can you explain what the space station uses its 'wings' for?"

"Wings?" Mike grinned, surprised. "Aren't those solar panels?"

"You're right," Dad said. "The ISS has solar panels, not wings. Together, these solar panels are big enough to cover six basketball courts. They are definitely the largest and most important part of the space station."

"Because they provide the space station with energy?" Mike asked.

FIGURE 15.6 The ISS above the Earth

"Exactly. The solar panels convert solar energy into electrical power for the entire space station. The communication networks, the experiments in the laboratories, the heating and cooling systems, and even the astronauts' supply of oxygen all depend on the electrical power generated by the solar panels. In a very real sense, the ISS is powered by the Sun.

"Of course, this shouldn't surprise us," Dad continued, "because every living thing on Earth (except for special, deep-sea bacteria) is also 'powered' by the Sun."

"You mean because plants use the energy in sunlight to produce food, and everything else in the world gets its energy from plants?" Mike asked.

"That's right," Dad replied. "When we were touring the biomes, you learned that plants and algae are primary producers. Through the process of **photosynthesis**, plants and algae use sunlight to convert non-living nutrients and minerals into food energy. When the plants or algae are eaten by primary consumers, and the primary consumers are eaten by other consumers, the solar energy that the plants store in sugars and carbohydrates is passed all the way up to the top predator."

"And it all goes back to the Sun's energy," Mike said, "just like on the ISS."

"Exactly," Dad said. "And that's why the next topic we're going to study is how the Sun and the other stars produce their energy."

"What's the difference between the Sun and a star?" Mike asked.

"There's no difference really," Dad answered. "Our Sun is a star like the other stars in the galaxy. The other stars have names like Betelgeuse and Sirius, and the star in our solar system is called the Sun."

"The Sun seems different from other stars because it is so important for us on Earth," Mike commented.

"That's true," Dad agreed. "When we consider how we depend on the Sun for the energy that provides us with food, light, and warmth, it makes sense that the Sun should seem more special than the other stars. Nevertheless, the Sun produces its energy in the same way as all the other stars, and it has the same life cycle as all the other stars of its size."

"Are we going to learn about those things?" Mike asked eagerly.

"Yes," Dad said. "We'll begin by learning how the Sun produces its energy."

---

**Photosynthesis** is the process by which plants and algae use sunlight to convert nutrients and minerals into food energy.

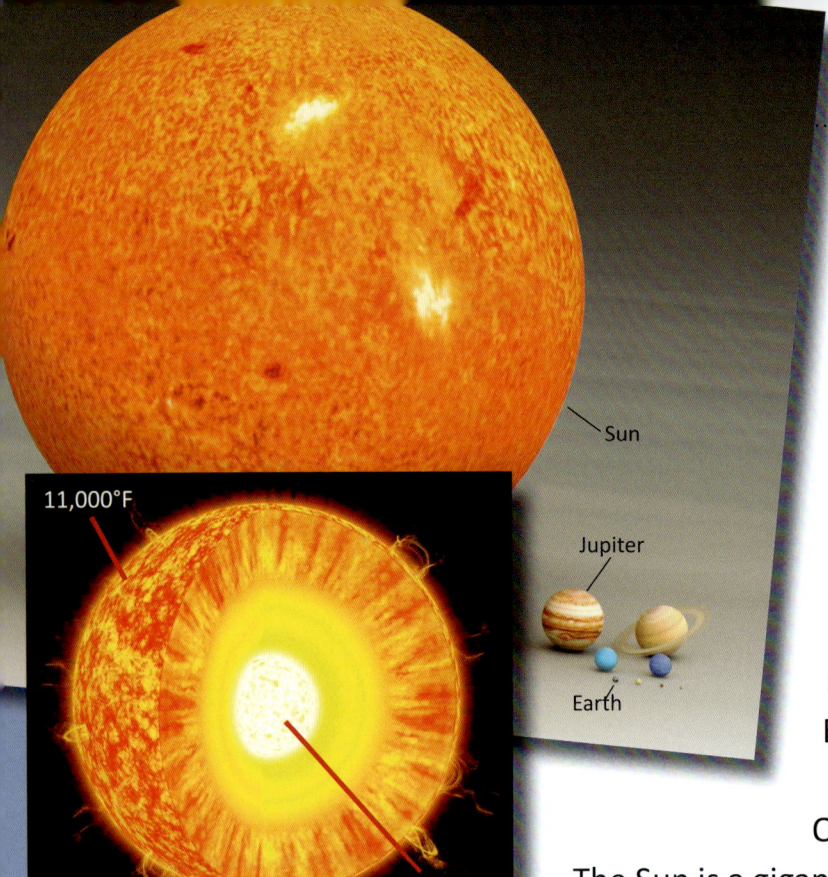

Figure 15.7

**The three states of matter** are the solid state, the liquid state, and the gaseous state.

**Nuclear fusion** (NOO-cleer FYOO-zhun) is a process in which two or more nuclei (NOO-klee-eye) are forced to **fuse**, or combine, into a single, larger nucleus. The process releases energy in the form of gamma rays.

## 15.4 Nuclear Fusion

The Sun is about 865,000 miles in diameter, which means it is approximately 110 times wider than the Earth. If the Sun were an eight-foot-high beach ball—big enough to reach from floor to ceiling in many homes—the Earth would be the size of a large marble. More than a million Earths could fit inside the Sun.

Of course, the Sun is not a beach ball. The Sun is a gigantic ball of hydrogen and helium gases. It is so hot that it radiates huge amounts of energy every second in the form of light and heat. The surface of the Sun is 11,000°F—hot enough to boil any metal. But the Sun's core is even hotter, at 27,000,000°F.

You learned about the **three states of matter** in Chapter 3, so you might be wondering how the Sun stays in the shape of a ball when it is not a solid. After all, gases do not usually have a definite size or shape. For instance, if someone opens the front door and lets a gust of cold air inside on a winter day, that air is not going to sit in a chunk on the doormat—it is going to spread through the room, making the whole house colder.

So how can the Sun stay together in the shape of a ball? The answer is gravity. As you have learned, gravity is a force that acts on matter, and the more matter an object has, the stronger its gravity is. The Sun has 333,000 times as much mass as the Earth, so its gravity is incredibly strong. The Sun's gravity is the reason that almost everything in the solar system orbits the Sun, and it is also the reason why the gases in the Sun do not turn into wisps of cloud and float away. The Sun's gravity is so strong that the gases in the Sun are held together in the shape of a ball by their own weight.

The Sun is almost one hundred times more massive than everything else in the solar system combined. Its enormous mass is what allows the Sun to produce its own energy. As you probably know, the Sun is the only thing in the solar system that shines by its own energy. All the other objects in the solar system—planets, moons, comets—shine because they are reflecting the light of the Sun. The Sun's incredible mass—greater than anything else in the solar system—turns it into an enormous furnace that produces light and heat through **nuclear fusion**.

Nuclear fusion occurs when two or more nuclei are forced to combine into a single, larger nucleus. You learned in Chapter 2 that an atom is made out of electrons, protons, and neutrons. The protons and neutrons are in the center of the atom, which is called the nucleus. The tiny electrons whizz around the nucleus on the outer edges of the atom.

When two nuclei slam into each other at tremendous speeds, they can combine to form a single, heavier nucleus. In the process, energy is released in the form of a **gamma ray**, which is a high-energy form of **electromagnetic radiation**.

Nuclei do not usually move at high enough speeds to fuse into a single nucleus. In fact, outside of science laboratories and

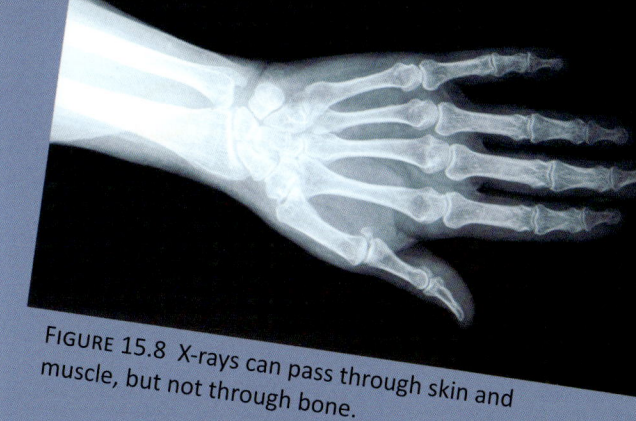

FIGURE 15.8 X-rays can pass through skin and muscle, but not through bone.

### Electromagnetic Radiation

As you remember from Chapter 4, **electromagnetic radiation is a way of transferring energy through electromagnetic waves.** X-rays, ultraviolet light, rays of sunshine, infrared rays, microwaves, and radio waves are all types of electromagnetic waves.

The difference between different types of electromagnetic waves is the amount of energy they carry. X-rays carry much more energy than rays of ordinary, visible light. That is why X-rays can pass through skin and muscle when visible light cannot. Radio waves carry much *less* energy than rays of visible light, so they are strong enough to carry sound and images to your television set, but not strong enough to provide you with light and heat.

**Gamma rays** are the most energetic, powerful form of electromagnetic radiation. Even though they are invisible, they carry much more energy than ordinary, visible light does. Gamma rays are so powerful that high doses of them are quite dangerous to living things.

experimental nuclear power plants, the only place in the solar system where nuclear fusion occurs is in the center of the Sun.

The Sun's center, or core, is the secret of its energy. Because of its tremendous mass, the Sun's gravity is incredibly strong. As a result, the Sun's inner core is literally crushed by the weight of its outer layers. The core is made largely out of hydrogen gas, but this hydrogen has been "squeezed" so powerfully by gravity that it is a million times denser than hydrogen on Earth. A cupful of the hydrogen gas in the Sun's core would weigh about 50 pounds on Earth!

Crushed inside the Sun's core, the hydrogen reaches temperatures of approximately 27,000,000°F. As you learned in Chapter 3, temperature is our way of measuring how much energy is possessed by the individual particles in an object. When the particles in an object have very little energy, they move and jiggle only a little, and we say they are cold. When the particles in an object have a lot of energy and are moving and jiggling a great deal, we say they are hot.

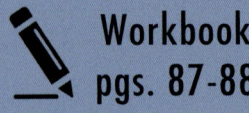
Workbook pgs. 87-88

The particles in an object are moving quickly even at 100°F, so imagine how quickly they are moving at 27,000,000°F! Nuclear fusion is possible in the Sun's core because the hydrogen is so hot that it is moving at incredibly fast speeds. The hot, "speedy" hydrogen nuclei in the Sun's core slam into each other with such force that every once in a while they stick together to become a heavier nucleus. When four hydrogen nuclei slam into each other in just the right order, the result is a helium nucleus. In the process, a little bit of energy is released in the form of two gamma rays. All the light and heat that we receive from the Sun started out as gamma rays released by colliding hydrogen nuclei.

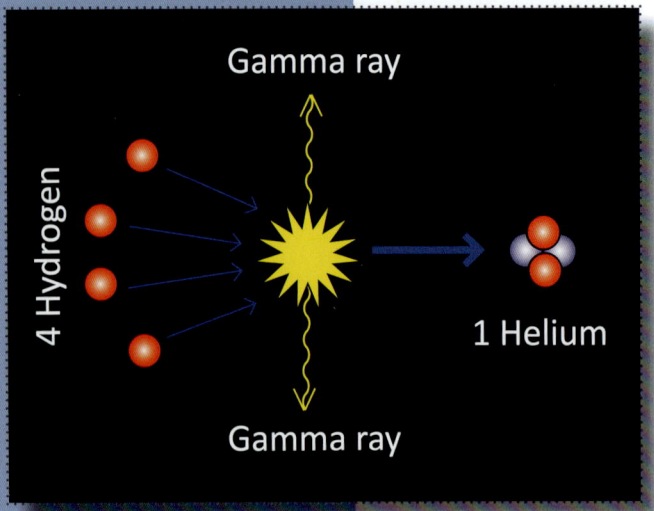
FIGURE 15.9 Hydrogen fusion

## 15.5 From Gamma Rays to Sunshine

"What do you think it would be like if four hydrogen nuclei fused into a helium nucleus right here in this room?" Dad asked Mike.

"There would be a huge explosion and our house would be blasted into outer space!" Mike exclaimed excitedly.

"Actually, no," Dad said. "We'd probably only hear a little 'pop,' if we even noticed the event at all. When hydrogen nuclei fuse into a helium nucleus, they only give off two gamma rays, and that's not much energy."

"It isn't?" Mike asked. "I thought gamma rays were the most powerful form of electromagnetic radiation."

"They are," Dad replied, "but even the most powerful form of electromagnetic radiation doesn't carry much energy in a single ray. It takes trillions and trillions of visible light rays to light a room, and it takes trillions and trillions of gamma rays to power the Sun. Even though a single gamma ray does not carry much energy, the Sun generates so many gamma rays that a single second's worth would be enough to meet the entire world's energy needs for the next 500,000 years.

"So let's imagine a gamma ray that has just been released by nuclear fusion in the center of the Sun. Like all other forms of light, gamma rays travel at the speed of light. So do you think it takes very long for the gamma ray to reach the Earth?"

FIGURE 15.10 One second's worth of the Sun's energy would be enough to meet the entire world's energy needs for the next 500,000 years.

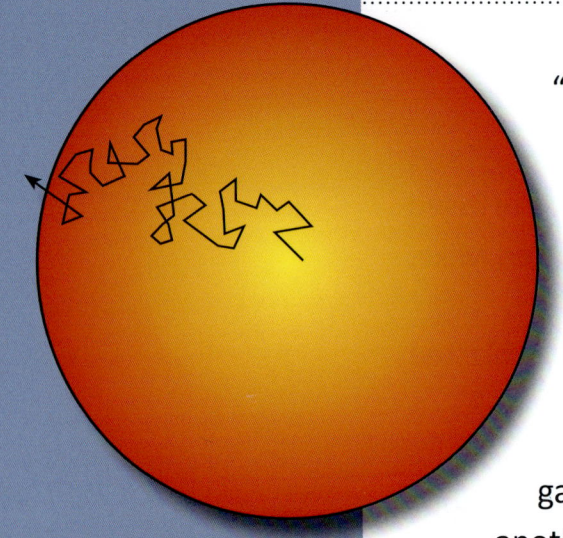

FIGURE 15.11 Path of a gamma ray travelling from the core to the surface of the Sun

Workbook pg. 89

"Nope," Mike said. "Light travels faster than anything else in the universe, so it takes hardly any time at all for the gamma ray to reach the Earth."

"That's what we'd expect," Dad agreed. "In reality, it takes a gamma ray hundreds of thousands of years just to travel from the core to the surface of the Sun. This is because the Sun is so dense. As soon as the gamma ray moves in one direction, it gets bumped in another direction, and ends up zigzagging all through the core (Figure 15.11). By the time the gamma ray reaches the Earth as a ray of sunlight, it is hundreds of thousands of years old."

"Boy! I'm glad I'm not a gamma ray waiting to get out of the Sun!" Mike grinned. "It sounds like being lost in a maze."

Dad laughed. "It might not be fun for the gamma rays, but it is definitely good for life on Earth. Gamma rays are so powerful that high levels of them are dangerous to human health. Thankfully, as the gamma rays make their way through the Sun, they lose some of their energy. Instead of emerging from the Sun as high-energy gamma rays, most of them emerge as visible light, or sunshine. And visible light is very good for human health, because it carries just the right amount of energy to power the process of photosynthesis in plants."

"And photosynthesis is the process that provides energy for the entire food chain," Mike concluded.

"Exactly," Dad agreed.

## 15.6 The Life Cycle of a Low-Mass Star

"Dad, the other evening you mentioned that our Sun has the same life cycle as other stars of its type," Mike said. "What do you mean by a life cycle?"

"That's a good question," Dad said. "The stars aren't actually alive, so we don't mean a life cycle like the life cycles of frogs and butterflies. But when a ball of gas begins producing energy through the process of nuclear fusion, scientists say that a star has been born, and when nuclear fusion ends in a star, scientists say that it has died."

"I didn't know that nuclear fusion in stars could end," Mike said. "I thought stars kept shining forever."

"Nuclear fusion continues for such a long time that it seems like forever to us," Dad admitted. "Even the most short-lived star shines for about 100,000 years. Some stars shine for one trillion years! But all stars eventually run out of fuel.

"As you know, the Sun shines because the hydrogen in its core is being fused into helium. The Sun has been fusing hydrogen into helium for about five billion years, and it will continue doing so for probably another five billion years. But eventually, all the hydrogen in the Sun's core will be used up and turned completely into helium. Without any hydrogen to fuse into helium, nuclear fusion will have to stop."

"And then the Sun will be dead?" Mike asked.

"Not yet. When nuclear fusion in a star stops, the amount of energy in the core decreases. Since the core of the star is no longer full of energetic gamma rays trying to force their way to the surface, gravity is able to compress the helium core even more.

**Low-Mass Star**

The hydrogen core

is fused into ...

a helium core

which . . .

is fused into ...

a carbon core.

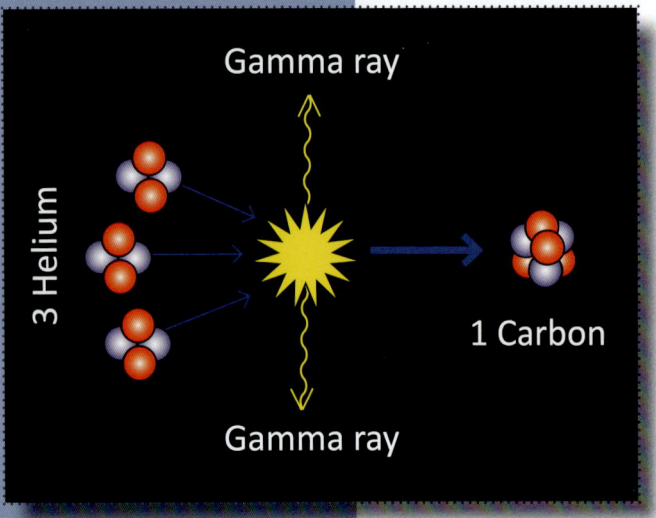

FIGURE 15.12 Helium fusion

A **planetary nebula** is the outer layers of a dead low-mass star.

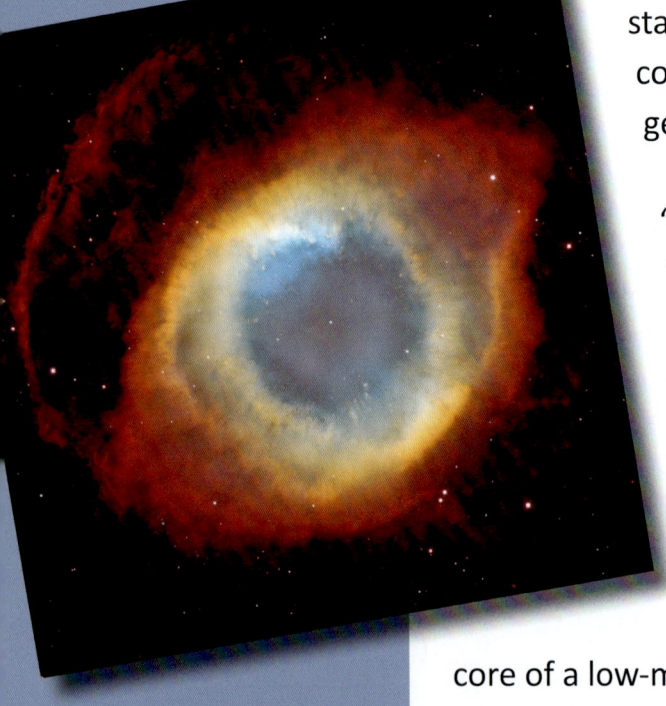

FIGURE 15.13 Helix Nebula

"As the helium in the core is compressed by the force of gravity, it becomes denser and denser and hotter and hotter. Eventually, the star's helium core reaches 180,000,000°F—hot enough for nuclear fusion to begin again. This time, hydrogen isn't fused into helium. Instead, helium is fused into carbon."

"I didn't know that was possible," Mike said. "Why wasn't helium fusing into carbon all along?"

"Because it takes much higher temperatures to fuse helium than it takes to fuse hydrogen. Helium fusion can't occur until a star has run out of hydrogen. Then gravity can compress the core farther to produce the higher temperatures needed to fuse helium."

"I guess the next question is what happens when the star runs out of helium," Mike said. "Does the gravity compress the core even more until the temperature gets high enough to fuse carbon?"

"Ah, this is where things get interesting," Dad said. "Different stars die in different ways. Some end up as dense, glowing diamonds called white dwarf stars. Others end with a bang in a supernova explosion.

"The factor that determines how a star ends its life is its mass, because the strength of a star's gravity depends on its mass. When the core of a low-mass star has been fused into carbon, its gravity is not strong enough to compress the core any farther. As a result, the carbon core of a low-mass star never gets hot enough to fuse into anything else. Instead, the outer layers of the star drift away in gigantic clouds of gas, called a **planetary nebula**."

Mike was puzzled. "Why are the clouds called a planetary nebula? Do they turn into planets or something?"

"Planetary nebulae got their name because the clouds of gas are often round, which makes a nebula look a bit like a planet through a telescope," Dad explained. "But planetary nebulae have nothing to do with planets—they are the outer layers of a dead star."

"What happens to the star's core once the outer layers float away?" Mike asked.

"Good question," Dad answered. "The carbon core of a low-mass star floats through space forever as a **white dwarf star**. A white dwarf star is a glowing ball of hot, dense carbon. It is called a star because its heat makes it shine, although not as brightly as real stars. The Sun is a low-mass star, so it will end up as a white dwarf."

"Oh, I was hoping the Sun would end with an explosion!" Mike said. "White dwarfs sound rather boring."

Dad smiled. "I know white dwarfs don't sound as exciting as giant explosions, but I like the idea of our Sun ending up as a huge, glowing diamond."

"What do you mean?" Mike asked. "White dwarfs are just balls of carbon, right?"

"So is a diamond," Dad replied. "Pencil lead and diamonds are both made out of carbon—the only difference is that diamonds are much, much denser than pencil lead."

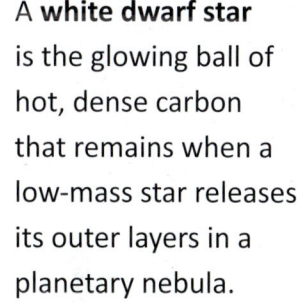

FIGURE 15.14 Eight-Burst Nebula

A **white dwarf star** is the glowing ball of hot, dense carbon that remains when a low-mass star releases its outer layers in a planetary nebula.

**Black Dwarf Stars**

A white dwarf star eventually cools so much that it no longer gives off light. Then it is called a black dwarf star.

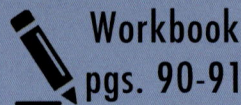

Workbook pgs. 90-91

**High-mass stars** are stars that are at least eight times as massive as the Sun.

A **supernova** (SOO-per-NOH-vuh) is the tremendous explosion that occurs at the death of a high-mass star.

302

"Whoa!" Mike exclaimed. "I never knew that! And the carbon in white dwarf stars is pretty dense, so it's probably more like a diamond than like a pencil lead, right?"

"The carbon in white dwarf stars is many times denser than anything we have on Earth," Dad said. "In fact, a cupful of white dwarf material on Earth would weigh the same as twelve elephants!"

"Diamonds and elephants," Mike said, shaking his head. "I've changed my mind about white dwarf stars. They're pretty amazing."

## 15.7  The Life Cycle of a High-Mass Star

"**High-mass stars** are stars that are at least eight times as massive as the Sun. High-mass stars end up as supernova explosions.

"At first, the death of a high-mass star and the death of a low-mass star are very much alike. For high-mass stars and low-mass stars, nuclear fusion stops when they run out of hydrogen fuel. Then the force of the star's gravity compresses the star's core, producing higher and higher temperatures until the helium begins to fuse into carbon.

"When the helium fuel is exhausted, nuclear fusion comes to a halt once more. Low-mass stars become white dwarf stars at this point, because their gravity is not strong enough to compress their carbon cores.

"Unlike low-mass stars, a high-mass star *does* have enough gravity to compress its carbon core. As the core is compressed, its temperature rises even further. Finally, the highly-compressed core reaches a temperature of one billion degrees, and the carbon in the core begins to fuse into neon or oxygen. The neon or oxygen is then fused into silicon, and the silicon is then fused into iron. By this time, the core is incredibly dense and has a temperature of nine billion degrees."

"Hold on a second—let me see if I have this right," Mike said. "In a high-mass star, hydrogen is fused into helium, which is fused into carbon, which is fused into neon or oxygen, which is fused into silicon, which is fused into iron. And this is all caused by the force of gravity, which raises the temperature of the core by pressing it into a smaller area. Is this correct?"

"Yes," Dad said.

"Then what does the iron fuse into?" Mike asked.

"Nothing," Dad said. "Iron can't be fused like hydrogen, helium, carbon, neon, oxygen, and silicon. Instead, the star becomes a **supernova**."

"Is a supernova a big explosion?" Mike asked, hopefully.

"A supernova is one of the most tremendous explosions in the universe," Dad said. "Frankly, I like to think of them as God's fireworks. A supernova explosion is about 700 septillion (700,000,000,000,000,000,000,000,000) times more powerful than the most powerful nuclear weapon, and can outshine an entire galaxy. When a supernova occurred in 1054 A.D., it was so bright that it was visible even during the daytime.

"After about a month, the supernova itself fades away, but the remnants of the exploded star are still there. Travelling outwards at high speeds, the glowing remnants of the star spread through space in delicate filaments and clouds. **Supernova remnants**, as they are called, can be seen for thousands of years after the supernova itself, and are incredibly beautiful."

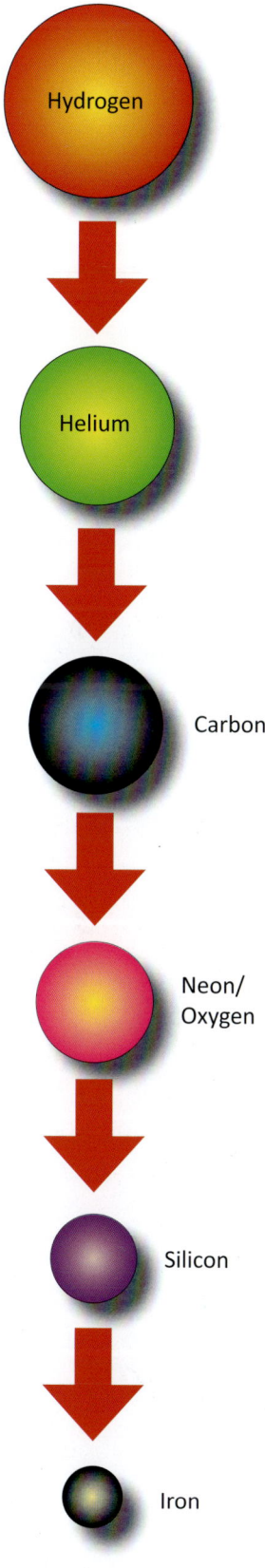

**High-Mass Star**

FIGURE 15.15 The Crab Nebula, the remnants of a supernova that occurred in 1054 A.D.

**Supernova remnants** are the glowing remnants of an exploded high-mass star.

## 15.8 Living on a Star

When a high-mass star becomes a supernova, it distributes throughout the galaxy the helium, carbon, neon, oxygen, silicon, and iron that has been fused in its core.

Of course, by the time a high-mass star dies, most of its hydrogen has been fused into helium, most of its helium has been fused into carbon, most of its carbon has been fused into neon or oxygen, and so on. But the star always ends up with a small layer of each element. When the supernova explodes, these elements are spread through the galaxy.

Supernovae also create completely new elements. A supernova explosion produces temperatures high enough to fuse elements such as gold, silver, cobalt, and nickel. The supernova distributes these newly-fused elements through the galaxy along with the elements that were fused inside the star. Except for hydrogen and helium, every element on Earth was formed either in the center of a star or in a supernova explosion.

"Among all the strange things that men have forgotten," G.K. Chesterton once wrote, "the most universal and catastrophic lapse of memory is that by which they have forgotten that they are living on a star." Chesterton was trying to ignite in his readers the same awe and appreciation for their home planet that we feel for stars, planets, comets, and other celestial bodies. After all, when the Earth is seen from outer space, it shines with all the brilliance of Venus, the brightest planet.

*Supernovae* (SOO-per-NOH-vee) is the plural of *supernova*.

Chesterton's statement is also literally true. We actually are living on a star, or at least the remnants of what was once a star. The atoms that form our mountains and oceans were forged millions of years ago deep inside a star or during a supernova. Plants, animals, and even our own bodies are made of elements that were fused in the gigantic furnaces of stars and supernovae. We and everything we see around us are literally made of star dust.

FIGURE 15.16 Earth from space

 Unit 3 Activity #6

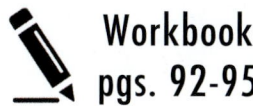 Workbook pgs. 92-95

# 16
## Chapter

# The Solar System

The Sun, with all those planets revolving around it and dependent on it, can still ripen a bunch of grapes as if it had nothing else in the Universe to do.
— Galileo Galilei

## 16.1 The Solar System

"Dad, what's that bright star above the western horizon?" Mike asked. "I can't find it on the star wheel at all, but it is even brighter than Sirius. Is it a supernova?" he asked hopefully.

"That is Venus," Dad said, "one of the eight **planets** in our solar system. I'm glad you noticed it, because we're going to be learning about the planets this week."

"Why don't they have Venus on the star wheel?" Mike asked. "It's the brightest star—I mean, thing—in the night sky."

"They don't have the Moon on the star wheel, either," Dad commented. "Why do you think that is?"

"Because the Moon moves around so much. It doesn't rise and set with the stars, so you can't chart it with the stars."

"The same thing applies to the planets. The movement of the planets through the sky is more complicated than the movement of the stars. This is because our view of the planets doesn't just depend on where the Earth is on its orbit around the Sun—it also depends on where the planets are in *their* orbit around the Sun. As the planets travel around the Sun on their different orbits, they move in and out of different constellations instead of remaining next to the same stars all year round."

"Then how do we know which 'stars' are really planets?" Mike asked.

> The eight **planets** are Mercury, Venus, Earth, Mars, Jupiter, Saturn, Uranus, and Neptune.

"There are lots of ways," Dad assured his son. "There are five planets that can be seen without a telescope, and any bright 'star' that isn't on the star wheel is almost certainly one of them. There are also many sources on the internet, in newspapers, and in astronomy magazines that provide up-to-date information on where the planets will be located each night. Monthly star charts sometimes indicate the general location of the planets during that month.

"With just a little experience, you can identify the planets without any help from the charts. As soon as you become familiar with the night sky for a particular time of year, you will start to notice that certain 'stars' wander among the other stars over time. Any bright 'star' that is near different stars every week is certainly a planet. In fact, that is where the planets got their name. The word 'planet' means 'wanderer' in Greek."

"So if I want to find a planet in the sky and I don't have a star chart with me, I have to look through the entire sky until I notice a star that was in a different place the previous week?" Mike asked, a little skeptical of this method.

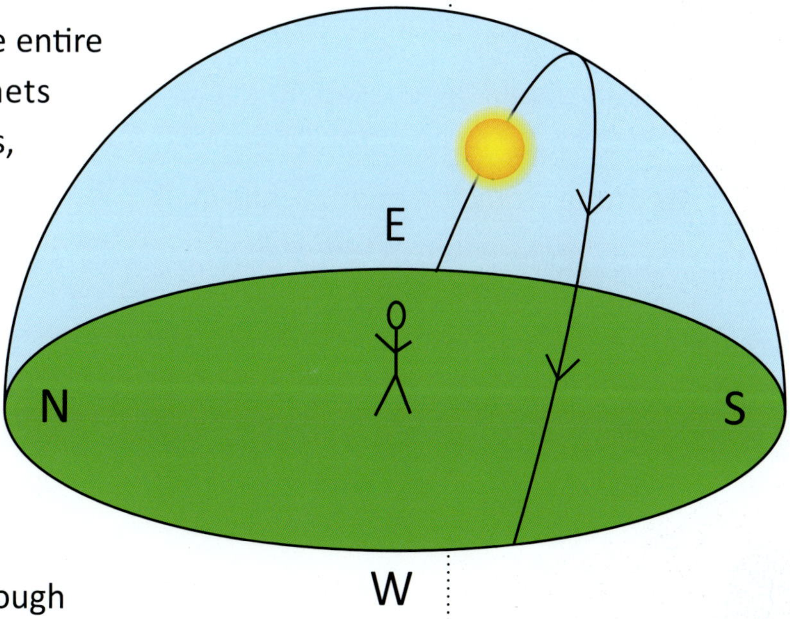

"You don't have to look through the entire sky," Dad assured him. "The planets move differently from the other stars, but they always stay near the path of the Sun through the sky. Of course, at night the Sun cannot be seen, but we can remember where it traveled across the sky in the daytime. The five planets that are visible without a telescope never 'wander' more than one fist above or below the Sun's path through the sky."

FIGURE 16.1 The path of the Sun through the sky

Mike was relieved that the planets stayed in a definite part of the sky. "Which are the five planets we can see without a telescope?" Mike asked. "Can we see them all now?"

"Mercury, Venus, Mars, Jupiter, and Saturn can all be seen without a telescope, so they are called naked-eye planets," Dad said. "They aren't usually visible at the same time, though. Right now I can see Venus, Jupiter, and Mars."

"Where? Show me!" Mike requested eagerly.

"Venus is the really bright one. It is the brightest object in the sky besides the Sun and Moon. It is often known as the Evening Star because it is the first 'star' to appear in the evening. Sometimes it appears in the morning instead of the evening, and then it is called the Morning Star.

"Mars is not always as bright as it is this year, but it is one of the only reddish objects in the sky besides Betelgeuse. This makes it fun to recognize.

---

**If the Sun were an 8-foot-high beach ball . . .**

**Mercury** would be a **pea**
*(diameter: 0.34 inch)*
**Venus** would be a **large marble**
*(diameter: 0.83 inch)*
**Earth** would be another **large marble**
*(diameter: 0.88 inch)*
**Mars** would be a **blueberry**
*(diameter: 0.47 inch)*
**Jupiter** would be a **basketball**
*(diameter: 9.86 inches)*
**Saturn** would be a **bowling ball**
*(diameter: 8.03 inches)*
**Uranus** would be a **baseball**
*(diameter: 3.24 inches)*
**Neptune** would be another **baseball**
*(diameter: 3.13 inches)*

Sun
Jupiter
Saturn
Uranus
Neptune
Earth
Venus
Mars
Mercury


FIGURE 16.2 Three-dimensional models of the planets and a slice of the Sun for size comparison

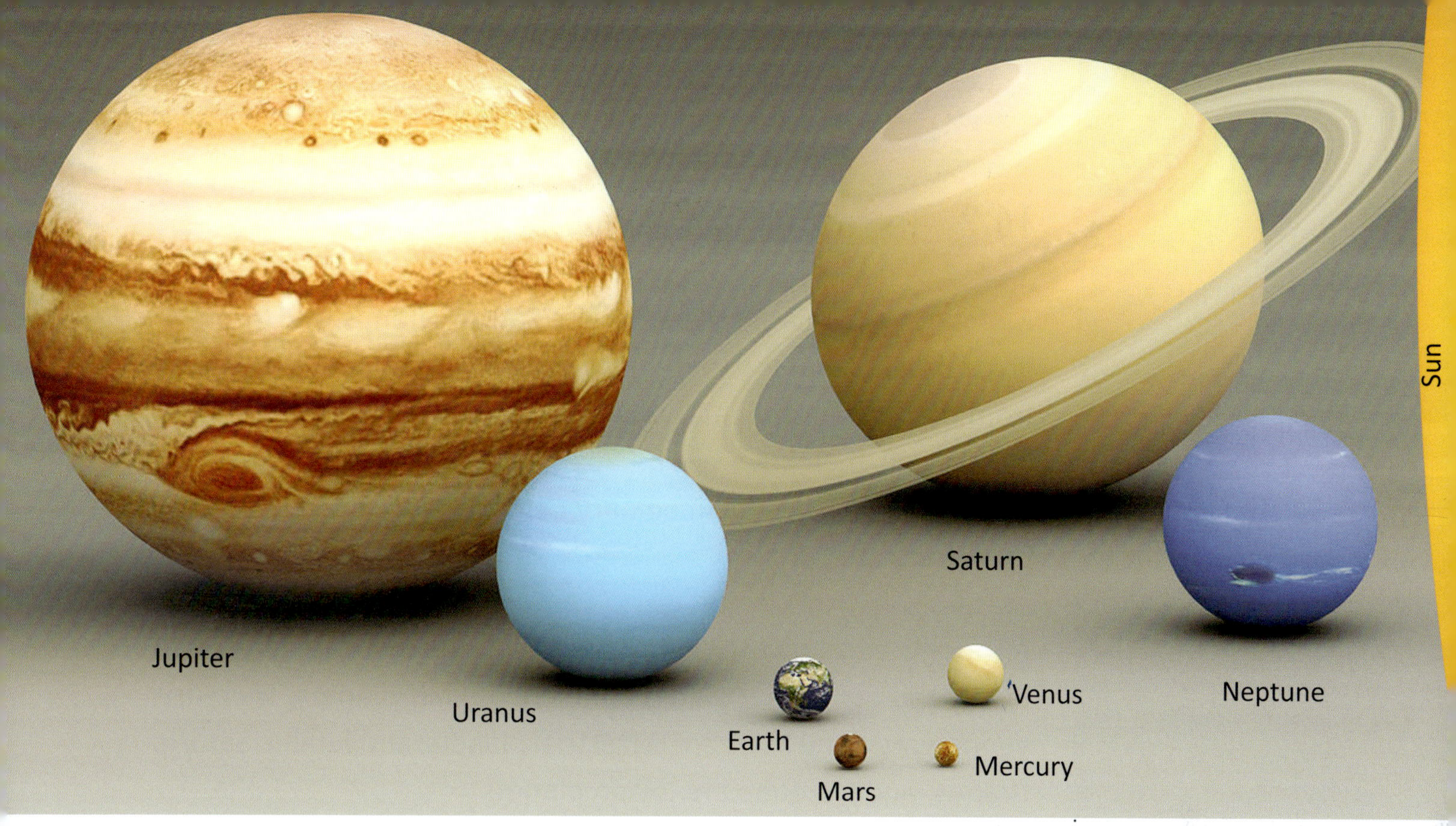

"Jupiter is brighter than Mars, but not as bright as Venus. Jupiter is special because it is possible to see some of its moons with a good pair of binoculars."

"Whoa! Jupiter has moons?" Mike exclaimed. "Let's get out the binoculars and look at them!"

"Sounds good to me," Dad said. "And during the next few days we'll learn more about the planets, including Jupiter and its moons."

FIGURE 16.3 If the **Sun** were an **eight-foot high beach ball**, the distances between the Sun and the planets would be . . .

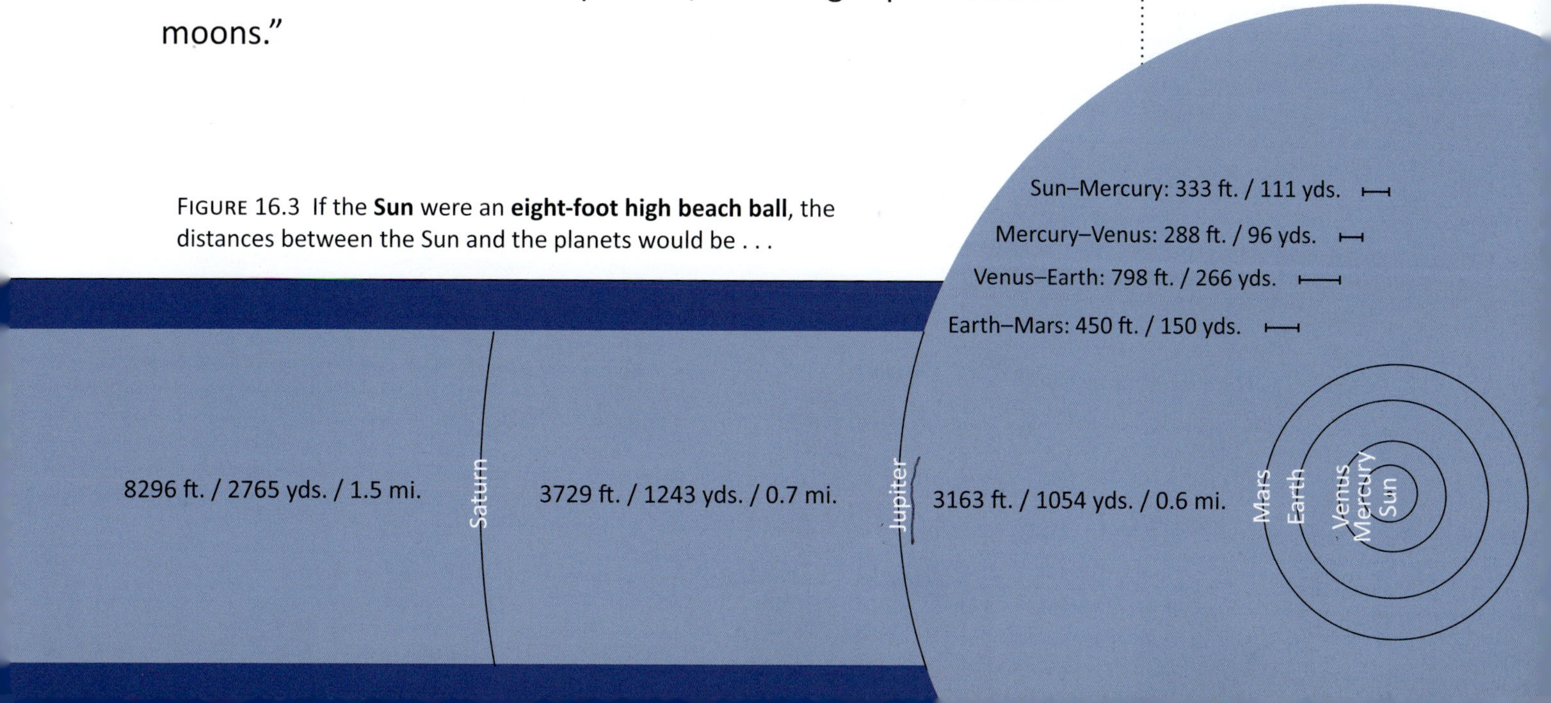

Sun–Mercury: 333 ft. / 111 yds.
Mercury–Venus: 288 ft. / 96 yds.
Venus–Earth: 798 ft. / 266 yds.
Earth–Mars: 450 ft. / 150 yds.
8296 ft. / 2765 yds. / 1.5 mi.
Saturn
3729 ft. / 1243 yds. / 0.7 mi.
Jupiter
3163 ft. / 1054 yds. / 0.6 mi.
Mars
Earth
Venus
Mercury
Sun

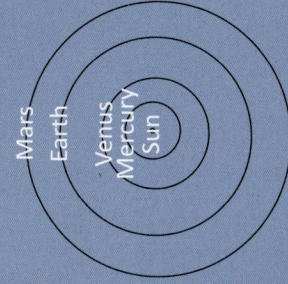

## 16.2 Terrestrial and Jovian Planets

The **terrestrial planets** (tuh-RES-tree-uhl PLAN-its) are made of rocks and metals, just like Earth.

The **jovian planets** (JOV-ee-an PLAN-its) are composed mainly of hydrogen and helium gases and are much larger than the terrestrial planets.

An **AU**, or **astronomical unit**, is the distance between the Earth and the Sun, about 93 million miles.

Mercury, Venus, Earth, Mars, Jupiter, Saturn, Uranus, and Neptune are the eight planets in the solar system, in order of their distance from the Sun. The planets revolve around the Sun and also rotate, or spin, on their own axes. ("Axes" is the plural of "axis." The axes of planets are the imaginary center lines around which the planets rotate.)

Although each of the planets is unique, there are two main types: the **terrestrial planets** and the **jovian planets**. Mercury, Venus, Earth, and Mars are terrestrial planets, and Jupiter, Saturn, Uranus, and Neptune are jovian planets.

The terrestrial, or Earth-like, planets get their name from the Latin word "terra," which means "earth." Unlike the huge jovian planets, which are composed mainly of hydrogen and helium gases, the terrestrial planets are made of rocks and metals, just like Earth. The terrestrial planets are all within two AU's of the Sun.

The jovian, or Jupiter-like, planets are composed mainly of hydrogen and helium gases and are much larger than the terrestrial planets. They are also much farther from the Sun. Jupiter, the jovian planet closest to Earth, is more than five AU's from the Sun, and Neptune is 30 AU's from the Sun. The jovian planets all have rings and moons, but since they are gaseous, they do not have solid surfaces. The jovian planets are often known as gas giants.

 Unit 3 Activity #7

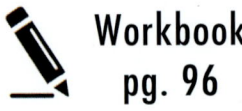 Workbook pg. 96

**Mercury** — First planet from Sun

| | |
|---|---|
| Average distance from Sun | 0.387 AU |
| Diameter | 3,032 mi. |
| Mass | 0.055 times Earth's mass |
| Length of year | 87.9 Earth days |
| Length of day | 58.6 Earth days |
| Surface temperature | Night: -280°F, Day: 800°F |
| Composition | Rocks, metals |
| Gravity | *100 lbs. on Earth = 38 lbs. on Mercury* |

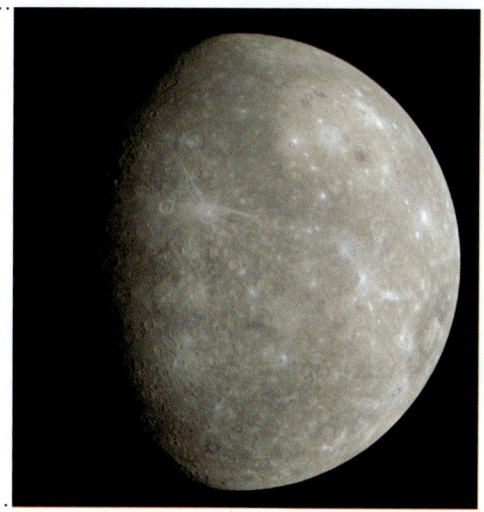

## 16.3 Mercury

Mercury is the smallest of the eight planets. Earth is more than twice as wide as Mercury, and is 18 times more massive. Because Mercury has so little mass, its gravity is quite weak. Because of Mercury's weak gravity, someone who weighs 100 pounds on Earth would weigh only 38 pounds on Mercury.

Mercury is also the planet closest to the Sun. The Earth is 93 million miles from the Sun, but Mercury is "only" 36 million miles from the Sun, less than half of the Earth-Sun distance. As you learned in Chapter 13, the Earth-Sun distance is often expressed as 1 AU (astronomical unit), so the distance between Mercury and the Sun can be expressed as 0.4 AU.

Since Mercury is so close to the Sun, we would expect it to be quite hot. Sure enough, Mercury is often 800°F during the day—as hot as glowing coals. At night, though, the temperature dips to -280°F— much colder than winter in Antarctica.

One reason for the planet's extreme temperatures is that Mercury rotates very slowly on its axis, so it takes much longer for the Sun to rise and set. A day on Mercury is 59 times as long as a day on Earth. The Sun blazes down on Mercury for 708 hours at a time (during Mercury's long days), and then is absent from the sky for the next 708 hours (during Mercury's long nights). No wonder Mercury's days are extremely hot and its nights extremely cold!

Mercury would not have such extreme temperatures if it had an atmosphere, though. The Sun shines down on Mercury during the day, warming its surface just as it warms the surface of the Earth. At night, the warm soil radiates the heat into outer space, leaving the night-time side of Mercury freezing cold. (As you learned in Chapter 4, everything that is warm radiates, or sends out, thermal energy through infrared radiation.)

On Earth, the heat of the Sun is not radiated back into space at night. This is because most of the infrared radiation (heat rays) that is radiated by the surface of the Earth is unable to pass through the Earth's atmosphere. Instead of escaping into outer space, the infrared rays are directed back towards the Earth. This is called the **greenhouse effect**, and it is caused by greenhouse gases such as carbon dioxide and water vapor. The greenhouse effect is the main reason Earth does not have extreme temperatures like Mercury.

The **greenhouse effect** is the result of greenhouse gases, which allow sunlight to pass through the atmosphere, but do not allow infrared radiation to pass back out.

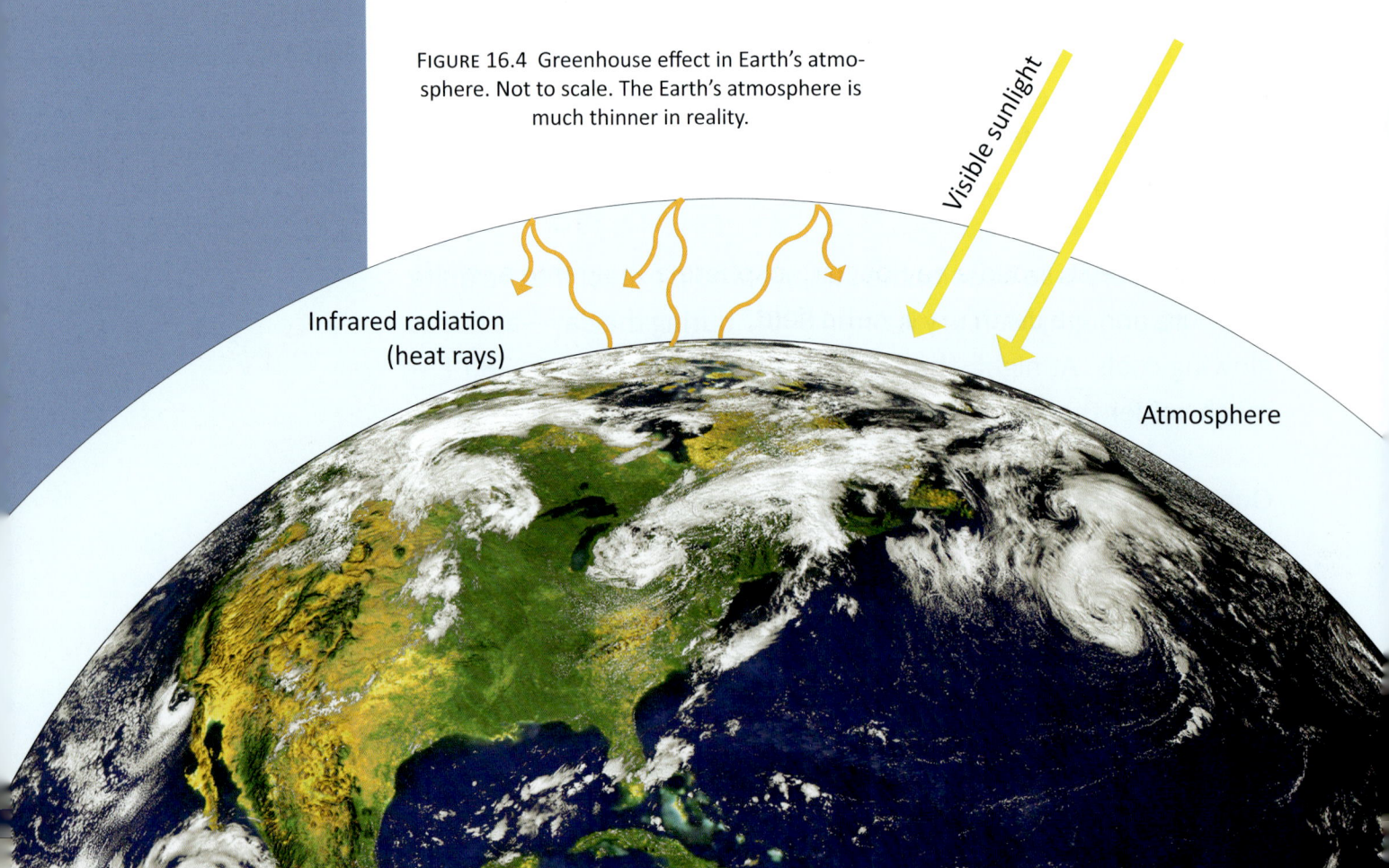

FIGURE 16.4 Greenhouse effect in Earth's atmosphere. Not to scale. The Earth's atmosphere is much thinner in reality.

The greenhouse effect can be seen on a smaller scale in greenhouses and parked cars. Like greenhouse gases in the atmosphere, glass allows rays of sunshine to pass through it, but blocks the passage of infrared radiation (heat rays). As a result, sunshine easily enters the greenhouse or the car, but the infrared radiation radiated by the warm flower beds or dashboards cannot escape back through the glass. Greenhouses and parked cars are essentially "heat traps," which is why the air inside greenhouses and parked cars is so much warmer than the air outside.

FIGURE 16.5

The fact that Mercury does not have an atmosphere explains why it experiences such temperature extremes. But what explains why Mercury does not have an atmosphere?

First of all, Mercury's gravitational pull is much weaker than the Earth's, so it is harder for it to "hold onto" an atmosphere. Any gases that originally covered Mercury would tend to drift away into space.

Mercury's loss of its atmosphere was also due to the solar wind. As you know, the Earth's magnetic field acts as a giant force field to protect our atmosphere from the solar wind. The solar wind would have stripped away our atmosphere a long time ago if it weren't for the Earth's magnetic field.

The Earth's magnetic field is caused by its molten iron core, which becomes a magnet when it is "sloshed around" by the motion of the Earth's rotation. Mercury also has a molten iron core, but the magnetic field it generates is very weak. In fact, Mercury's magnetic field is only one-hundredth as strong as Earth's magnetic field. Such a weak magnetic field is not enough to protect Mercury's atmosphere from being stripped away by the solar wind.

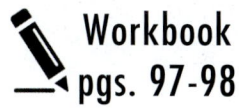

**Workbook pgs. 97-98**

**Venus** — Second planet from Sun

| | |
|---|---|
| Average distance from Sun | 0.723 AU |
| Diameter | 7,520 mi. |
| Mass | 0.815 times Earth's mass |
| Length of year | 225 Earth days |
| Length of day | 243 Earth days |
| Surface temperature | 864°F |
| Composition | Rocks, metals |
| Gravity | 100 lbs. on Earth = 91 lbs. on Venus |

## 16.4 Venus

"Do you think you'd like to visit Venus?" Dad asked Mike.

"Definitely. It is the brightest, most beautiful planet in the sky," Mike said.

"True, but appearances can be deceiving. Venus' beauty comes from its thick atmosphere and clouds, which reflect a great deal of sunlight, making Venus the brightest object in the sky second to the Sun and the Moon. But Venus' atmosphere is precisely what makes it one of the most inhospitable of the terrestrial planets."

"With Mercury, it was *not* having an atmosphere that made it an unpleasant place to live," Mike said. "How can having an atmosphere make Venus inhospitable?"

"Have you ever heard of 'too much of a good thing'?" Dad asked. "Venus' atmosphere is so dense that it is almost like a liquid. In fact, Venus' atmosphere is so thick that the **atmospheric pressure** is 90 times greater than the atmospheric pressure on Earth."

"What is atmospheric pressure?" Mike asked.

"It's the way we measure the weight of the air above our heads and the 'push' of the air against our bodies. We don't usually notice the pressure of the air around us, but it is there, just the

**Atmospheric pressure** is the way we measure the weight of the air above our heads and the "push" of the air against our bodies.

same. If we swim underwater, water presses on us in the same way. Of course, water pressure is more noticeable than atmospheric pressure, because water is denser and heavier than air."

"Is water pressure the reason my ears sometimes hurt when I dive down to the deep end of the swimming pool?" Mike asked.

"Yes," Dad said. "When you dive beneath the surface, the pressure you feel in your ears is caused by the pressure of the water above and around you. The farther you go beneath the surface, the greater the pressure becomes because there is more water pressing down on you. At only 33 feet below the surface, the volume of your lungs would be compressed to half their normal size by the force of the water pressing on your body."

"Wow! I'm glad the pressure of the atmosphere isn't that strong!" Mike said.

"The atmospheric pressure on Earth isn't that strong, but Venus' atmospheric pressure is much stronger. To feel a similar pressure on Earth you would have to dive half a mile beneath the surface of the ocean—something no human being can do without special equipment to keep his lungs from being compressed."

"Umm, I change my mind about wanting to visit Venus," Mike said.

"I'm glad," Dad laughed, "because the intense atmospheric pressure is only one of the reasons Venus would be an uncomfortable planet to visit. Venus' thick atmosphere also causes a runaway greenhouse effect. As you learned when we

FIGURE 16.6 An atmospheric diving suit allows divers to descend into very deep water.

studied Mercury, the greenhouse effect is caused by greenhouse gases. The Earth's atmosphere contains greenhouse gases—especially carbon dioxide, water vapor, and methane—but they are less than one percent of the atmosphere. Because of this one percent, Earth's atmosphere 'traps' just the right amount of heat during the night and allows the rest of it to escape into space.

"Unlike Earth's atmosphere, Venus' dense atmosphere is 97% carbon dioxide, one of the most important greenhouse gases. Venus' carbon-dioxide atmosphere traps too much heat! The surface temperature on Venus is 864°F, even at night."

"Yes, my trip to Venus is definitely off," Mike declared.

"Good," Dad said. "Even the unmanned space probes that landed on the surface of Venus in the 1970's couldn't survive the heat and pressure longer than two hours. In fact, I think we should move on to the next planet before we undergo the same fate as those space probes!"

## 16.5  Mars

Mars, the red planet, is similar to Earth in many ways. The length of a Martian day is only a little longer than an Earth day, although a Martian year—the time it takes Mars to revolve around the Sun—is nearly twice as long as an Earth year. Because Mars' axis is tilted like the Earth's, it experiences four seasons just as the Earth does. Like Earth and unlike the other terrestrial planets, Mars is orbited by a moon—in fact, by two moons. The Martian moons are fairly small and are named Phobos and Deimos.

Mars is not much bigger than Mercury, so it is much smaller than the Earth. However, Mars contains one of the tallest mountains in the solar system: Olympus Mons. Olympus Mons is 16 miles high, which makes it three times taller than Mount Everest, the tallest

*The War of the Worlds*, an 1898 novel by H.G. Wells, describes a fictional invasion of Earth by aliens from Mars. When a radio adaptation of *The War of the Worlds* was performed in 1938 on Orson Welles' Mercury Theatre on the Air, many listeners panicked, believing an invasion was actually taking place.

**Mars** — Fourth planet from Sun

| | |
|---|---|
| Average distance from Sun | 1.52 AU |
| Diameter | 4,212 mi. |
| Mass | 0.107 times Earth's mass |
| Length of year | 1.88 Earth years |
| Length of day | 24.6 hours |
| Surface temperature | Night: -125°F, Day: 23°F |
| Composition | Rocks, metals |
| Moons | 2 |
| Gravity | *100 lbs. on Earth = 38 lbs. on Mars* |

mountain on Earth. Is it just chance that Mars has a mountain three times larger than the tallest mountain on Earth, even though Mars itself is much smaller than Earth? Not at all. The larger a planet is, the smaller its mountains have to be. This is because the gravity of a massive planet is stronger than the gravity of a less massive planet.

Imagine moving Olympus Mons from Mars to the Earth. Although Olympus Mons was certainly heavy on Mars, it is more than twice as heavy on Earth, because Earth's gravity is stronger than the gravity of Mars. When you set Olympus Mons down, perhaps in the middle of the United States, the weight of the mountain shakes the ground and presses the mountain deep into the Earth's crust.

As Olympus Mons rests on the crust of the Earth, the weight of the mountain exerts great pressure on the base of the mountain. As the pressure continues, the base of Olympus Mons becomes extremely hot, and starts to melt. Of course, as the mountain melts, it also sinks into the crust, becoming shorter and shorter. Eventually, the mountain will stop sinking, but not until it is about the height of Mt. Everest, which is the highest

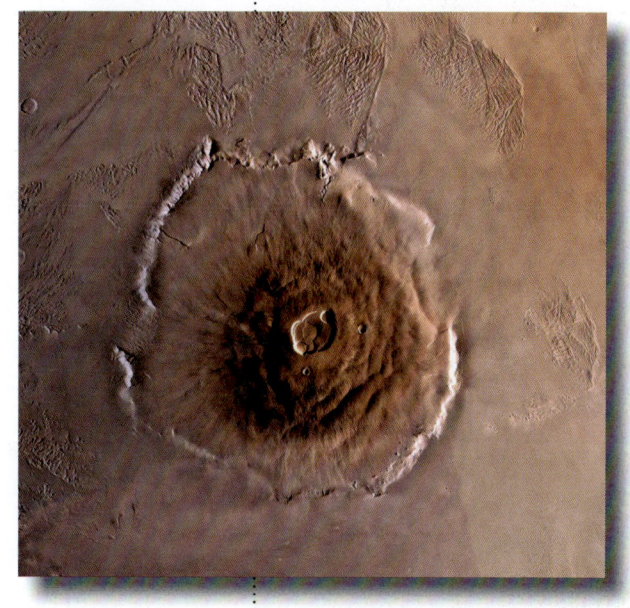

FIGURE 16.7 Olympus Mons

an Earth-mountain can be without melting. The reason Olympus Mons does not melt on Mars is that the gravity of Mars—and therefore the weight of Olympus Mons—is so much less than the gravity of the Earth.

Mars is the only terrestrial planet besides Earth that contains significant amounts of water. All the water on Mars is in a frozen state, however, because the temperature on Mars is usually well below the freezing point. Mars has an average temperature of about -60°F, and at its north and south poles, the temperature can fall to -200°F. This is cold enough to freeze carbon dioxide into dry ice! The polar ice caps on Mars are made both of frozen water and frozen carbon dioxide.

One reason for the cold temperatures on Mars is that Mars is 1.52 AU's from the Sun, or one and a half times as far from the Sun as the Earth is. But the real reason that Mars is so cold is that it has a very thin atmosphere. Because Mars is quite small, its molten iron core cooled and hardened relatively quickly, so it lost the magnetic field that had been protecting its atmosphere from the solar wind. Just like Mercury, Mars has lost most of its atmosphere to the solar wind.

If Mars had a thicker atmosphere, the greenhouse effect could trap enough of the Sun's heat to maintain comfortable temperatures, even though Mars is so far away. In fact, it seems certain that Mars did have warmer temperatures in the past, perhaps before it lost its atmosphere because of the cooling of its iron core. The evidence for warmer temperatures on Mars comes from images of the Martian surface that were taken by satellites orbiting Mars. Some of the craters on Mars have been **eroded** as if by rainfall, and there are channels that look like dry riverbeds. Some of the

To **erode** (ee-ROAD) means to wear away. Erosion is the wearing away of the land.

FIGURE 16.8 Eroded channels in a crater in the southern highlands of Mars. The surface of Mars is red because it contains iron oxide, or rust.

rocks on Mars were formed by sedimentation, a process which involves the hardening of sand at the bottoms of rivers and lakes.

As far as scientists know, these types of erosion cannot occur without flowing water. Thus, there must have been a time when Mars was warm enough to have large amounts of liquid water. Indeed, some astronomers believe liquid water might still flow occasionally, warmed by the remaining heat in Mars' core. But most of the water on Mars probably escaped into space when Mars lost its magnetic field. Despite the similarities between Mars and the Earth, there truly is no place like home.

## 16.6 The Asteroid Belt and Jupiter

There is a gap of almost four AU's between Mars, the farthest terrestrial planet, and Jupiter, the nearest jovian planet. This gap is not empty space; rather, it contains the **asteroid belt**, which contains most of the asteroids in the solar system.

**Asteroids** are large chunks of rock and metal that orbit the Sun. As you can see in Figure 16.9, most asteroids have very irregular shapes and look more like potatoes than like round balls. This is because they are much less massive than planets, so their gravity is not strong enough to pull them into round, spherical shapes.

There are millions of asteroids in the asteroid belt. Some are only a few yards across and others are more than 100,000 miles across. Although none of the asteroids is as large as a planet, some are large enough to have small moons. The largest asteroid in the asteroid belt is named Ceres. It is so large that

The **asteroid belt** (AS-tuh-royd belt) is the gap between Mars and Jupiter that contains most of the asteroids in the solar system.

**Asteroids** (AS-tuh-roydz) are large chunks of rock and metal that orbit the Sun.

 Experiment #26

FIGURE 16.9 Asteroids of various sizes

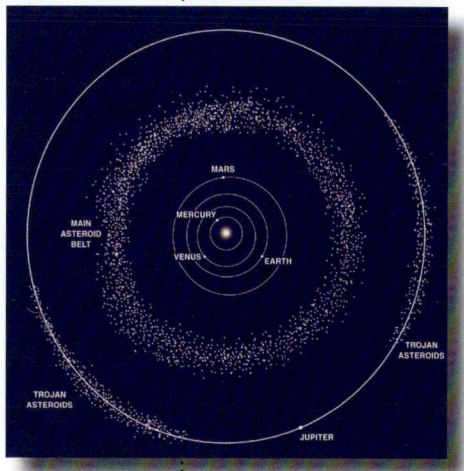

FIGURE 16.10 The asteroid belt

FIGURE 16.11 Ceres, the largest asteroid in the asteroid belt

it is sometimes known as a minor planet. As you can see in Figure 16.11, its gravity has even pulled it into the shape of a sphere.

Millions of asteroids might sound like a lot, but it is not very many considering how large the solar system is. On average, the asteroids in the asteroid belt are thousands of miles apart. The scarcity of asteroids in the solar system is crucially important for the Earth. Imagine what would happen if there were asteroids near the Earth! Even a small asteroid could cause great damage if it collided with the Earth, so we are very lucky that our path around the Sun is almost completely clear of asteroids.

The scarcity of asteroids in our part of the solar system is not due to luck, however. Jupiter, the jovian planet closest to Earth, acts as a giant "sweeper" to clear away asteroids from the inner solar system.

Ten times wider than the Earth and 318 times more massive, Jupiter is the largest object in the solar system after the Sun. Because of its enormous mass, Jupiter has an extremely strong gravitational pull—a person weighing 100 pounds on Earth would weigh 253 pounds on Jupiter. Scientists have discovered that

FIGURE 16.12 Arizona's Meteor Crater was formed when a relatively small asteroid (approximately 80 feet wide) collided with the Earth.

even the Sun wobbles a little because of the strength of Jupiter's gravitational pull.

Because Jupiter's gravity is so strong, asteroids millions of miles away are affected by its pull. Jupiter's gravity is so strong that it has swept most of the asteroids from the inner part of the solar system out of their original orbits. Many were pulled directly towards Jupiter and were absorbed into its mass. Other asteroids received only a swift gravitational tug as Jupiter sped past them on its path around the Sun. These asteroids collided with other planets or were flung to the outermost regions of the solar system. Thanks to Jupiter's work as a "sweeper planet," our part of the solar system is almost entirely free of asteroids.

**Jupiter** — Fifth planet from Sun

| | |
|---|---|
| Average distance from Sun | 5.2 AU |
| Diameter | 86,880 mi. |
| Mass | 318 times Earth's mass |
| Length of year | 11.9 Earth years |
| Length of day | 9.93 hours |
| Surface temperature | -234°F |
| Composition | Hydrogen, helium |
| Known moons | 67 |
| Gravity | *100 lbs. on Earth = 253 lbs. on Jupiter* |

## 16.7  More about Jupiter

Jupiter's orbit is quite far from the Sun, so its surface is extremely cold: -234°F. Of course, Jupiter does not have a solid surface, because it is formed almost entirely of hydrogen and helium gases.

If Jupiter were about 80 times more massive, the crushing pressure on its inner regions would be great enough for nuclear fusion to begin. In other words, Jupiter would be a small star, producing its own energy through nuclear fusion.

When space probes (unmanned, of course) approach Jupiter, they do not land on a rocky surface; rather, they plunge into thicker and thicker clouds of gas.

Even though Jupiter is composed of hydrogen and helium, the inner regions of the planet are so dense that the hydrogen and helium act more like liquids and solids than like gases. This is the effect of Jupiter's gravity, which pulls every part of the planet towards the center with tremendous force. Just as the center of the Sun is crushed by the weight of its outer layers, so the center of Jupiter is crushed by *its* outer layers. Since Jupiter is much less massive than the Sun, though, its inner regions are not crushed enough for nuclear fusion to begin. This is why Jupiter counts as a planet and not as a star.

Like all the jovian planets, Jupiter rotates extremely quickly. Even though Jupiter is 10 times as wide as the Earth, it completes a full rotation in less than 10 hours. Jupiter's stripes are the result of strong east-west winds caused by the planet's fast rotation. Wind speeds of more than 250 miles per hour on Jupiter make a 150-mph hurricane seem like a gentle breeze! Jupiter's Great Red Spot is a huge storm that has lasted for at least 300 years and is large enough to swallow two or three Earths.

Unlike the terrestrial planets, Jupiter is orbited by dozens of large and small moons. Besides its four larger moons— Callisto, Ganymede, Europa, and Io—Jupiter has at least sixty smaller moons as well. Ganymede is about the size of Mercury, which makes it the largest moon in the solar system.

Jupiter's four larger moons are known as the Galilean moons because they were first discovered by Galileo Galilei.

FIGURE 16.13 Jupiter's Great Red Spot

FIGURE 16.14 Jupiter's four largest moons: Callisto, Ganymede, Europa, and Io

The Galilean moons are made of rock, metal, and ice. Callisto, Ganymede, and Europa are covered in ice, but they may have deep oceans of liquid water beneath the surface. The icy surface of Europa is especially likely to cover an ocean of liquid water.

Unlike the other three Galilean moons, Io is covered with active volcanoes instead of ice. Since Io is much smaller than Mercury, we would expect it to have lost its interior heat a long time ago, just as Mercury and our own Moon did. On the contrary, Io has so much interior heat that it is the most volcanically active place in the solar system. Astronomers now know that Io's interior heat comes from tidal heating, a process in which Io is stretched and pulled in different directions by the strength of Jupiter's gravity. The interior friction caused by this stretching and pulling—much like the friction that heats silly putty when it is twisted and stretched—is the reason Io has so many volcanoes.

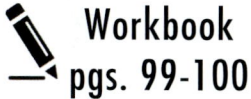

**Workbook pgs. 99-100**

## 16.8 Saturn

Saturn is the second jovian planet from the Sun and the second largest planet in the solar system. Saturn is much like Jupiter in many ways. Since Saturn is made mostly of hydrogen and helium gases, it does not have a solid surface. Saturn rotates very quickly on its axis, and thus experiences very strong winds. Saturn has over 60 moons, including Titan, one of the largest moons in the solar system.

**Saturn** — Sixth planet from Sun

| | |
|---|---|
| Average distance from Sun | 9.54 AU |
| Diameter | 72,367 mi. |
| Mass | 95 times Earth's mass |
| Length of year | 29.4 Earth years |
| Length of day | 10.6 Earth hours |
| Surface temperature | -288°F |
| Composition | Hydrogen, helium |
| Known moons | 62 |
| Gravity | 100 lbs. on Earth = 107 lbs. on Saturn |

Saturn is most famous for its wide system of rings. The other jovian planets also have rings, but Saturn's are the largest and most impressive. Saturn's rings look like solid sheets of shining "ring material," but in reality they consist of millions of ice particles all revolving around Saturn. The icy particles can be as small as a grain of dust or as large as an iceberg. The icy particles reflect the light of the Sun so well that it is possible to see Saturn's rings from one's backyard with a small telescope.

## 16.9  Uranus and Neptune

Uranus and Neptune are the farthest planets from the Sun. They are smaller than Jupiter and Saturn, but still much larger than Earth.

**Uranus** (YOOR-a-nus) — Seventh planet from Sun

| | |
|---|---|
| Average distance from Sun | 19.2 AU |
| Diameter | 31,518 mi. |
| Mass | 14.5 times Earth's mass |
| Length of year | 83.8 Earth years |
| Length of day | 17.2 Earth hours |
| Surface temperature | -357°F |
| Composition | Hydrogen, helium, methane |
| Known moons | 27 |
| Gravity | 100 lbs. on Earth = 91 lbs. on Uranus |

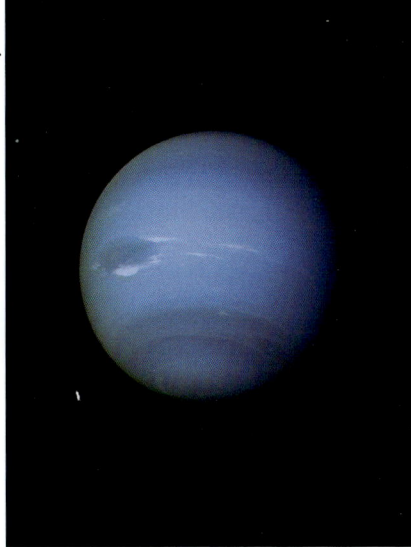

**Neptune** — Eighth planet from Sun

| | |
|---|---|
| Average distance from Sun | 30.1 AU |
| Diameter | 30,598 mi. |
| Mass | 17.2 times Earth's mass |
| Length of year | 165 Earth years |
| Length of day | 16 Earth hours |
| Surface temperature | -353°F |
| Composition | Hydrogen, helium, methane |
| Known moons | 14 |
| Gravity | 100 lbs. on Earth = 114 lbs. on Neptune |

Uranus and Neptune are gas planets, so they do not have solid surfaces. Unlike Jupiter and Saturn, Uranus and Neptune contain large amounts of water, ammonia, and methane in addition to hydrogen and helium. The methane in their atmospheres is what gives them their blue-green color.

The most extraordinary feature of Uranus is the tilt of its axis. Compared to the rest of the solar system, Uranus is tipped on its side by 98°. While the other planets orbit the Sun like spinning tops, Uranus travels along its orbit like a rolling ball. As a result, the Sun does not rise and set on Uranus the way it does on other planets. Instead, the Sun slowly rises and sets once every time Uranus travels around the Sun. It takes Uranus 165 Earth-years to orbit the Sun once, so if we lived at the north pole of Uranus, our days and nights would each be 82.5 Earth-years long.

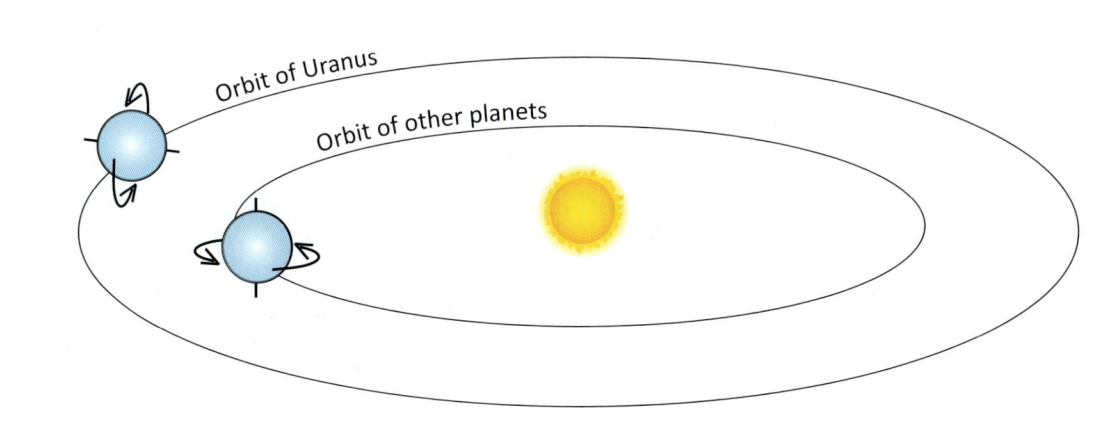

FIGURE 16.15 Uranus travels along its orbit like a rolling ball. Not drawn to scale

## Pluto and Kuiper Belt Objects

Beyond the orbit of Neptune, millions of small, icy objects travel around the Sun. These objects form the **Kuiper Belt**, a ring of comets and dwarf planets on the outskirts of the solar system.

Until 2006, **Pluto** was the ninth planet in the solar system. It is now considered a dwarf planet and one of the largest Kuiper Belt objects. Twenty-five times smaller than Mercury and only two-thirds the size of the Earth's Moon, Pluto is an object of ice and rock with five known moons.

FIGURE 16.16
Pluto and its largest moon Charon

The **Kuiper Belt** (KYE-per belt) is a ring of comets and dwarf planets on the outskirts of the solar system.

**Pluto** was once the ninth planet in the solar system. It is now considered a dwarf planet and one of the largest Kuiper Belt objects.

**Comets** are huge chunks of ice mixed with rocks and dust.

## 16.10 Comets and Meteor Showers

When you hear the word "comet," you probably think of a star with a long, shining tail. This is what comets look like from Earth, but **comets** are actually huge chunks of ice mixed with rocks and dust. Most comets are hundreds of AU's away and completely invisible from the Earth. Once in a while, though, a comet enters the inner solar system.

When a comet approaches the Sun, its ice begins to sublimate[1] into vapor because of the warmth of the Sun's rays. The vapor that is released includes the tiny particles of dust that were mixed with the ice. The gas and dust form a huge cloud that stretches behind the comet for millions of miles.

In photos, comets often look like they are speeding through the sky like shooting stars. In reality, they move through the night sky as slowly and gracefully as the stars and planets. The shining tails of comets are not caused by their high speeds as they orbit

---

[1] As you learned in Chapter 3, sublimation occurs when a solid changes directly into a gas without first going through the liquid state.

the Sun. Rather, the cloud of dust and gas that surrounds a comet is blown into a tail by the outward pressure from the Sun and the solar wind. A comet's tail always flows away from the Sun because that is the direction in which it is being blown.

When a comet appears in the night sky, it is usually visible for several weeks. During this time, the main body of the comet and its huge tail shine because they are reflecting the Sun's light. When the comet's orbit takes it back into the freezing outer regions of the solar system, its tail disappears and the comet itself fades from sight. Travelling farther and farther away, the comet rejoins its fellow comets on the outer edges of the solar system. There, it follows its distant orbit around the Sun as a rocky, dusty chunk of ice.

FIGURE 16.17 Comet Lovejoy above the Earth's horizon, as seen from the International Space Station. In which direction is the Sun?

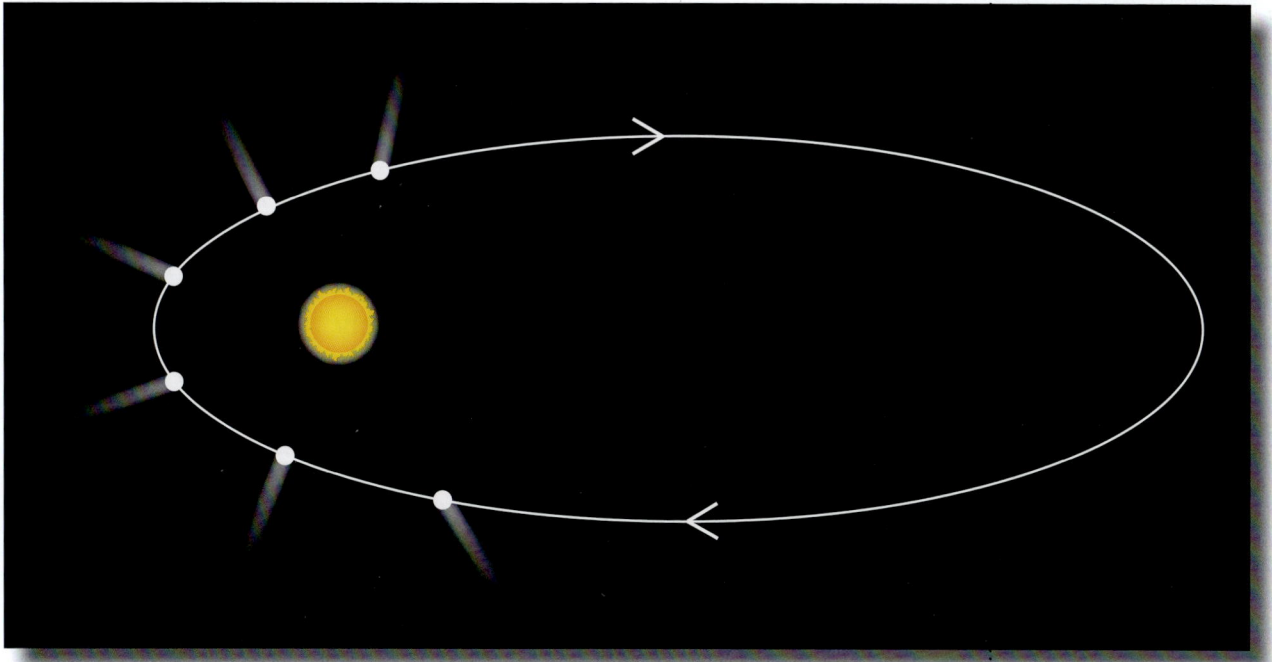

FIGURE 16.18 Orbit of a comet

FIGURE 16.19  Time-lapse photo of meteor shower

**Meteors** (MEE-tee-erz) are rocky particles that fall through the Earth's atmosphere as shooting stars.

Even though comets do not remain visible for more than a few weeks, they leave behind them a trail of rocky particles about the size of sand or pebbles. At certain times of the year, the Earth encounters these particles on its orbit around the Sun. When the particles enter the Earth's atmosphere, they become glowing **meteors**—or as most people call them, shooting or falling stars. Meteors glow because the friction between the meteors and the air heats the meteors to extremely high temperatures. Most meteors are vaporized completely before they reach the Earth, disappearing in streaks of light. Only a few meteors ever survive the trip through the Earth's atmosphere and strike the Earth. The few meteors that do strike the Earth are usually too small to do any damage, and no one has ever been reported as being injured.

When the Earth passes through the debris of a comet, so many meteors can be seen that the event is known as a **meteor shower**. The best meteor showers can produce 60 meteors per hour. Meteor showers occur at the same times each year because Earth collides with the debris of long-gone comets at the same places each year. Of course, there are many types of debris scattered through the solar system, and not all of it was left by comets. There is enough random debris along Earth's orbit that several meteors per hour can be seen on any clear, moonless night.

**Meteor showers** occur when the Earth passes through the debris of a comet.

Since the Earth collides with the debris of comets at the same places in its orbit each year, it is possible to predict when meteor showers will occur. The table below lists the major meteor showers and the dates and times when they can be viewed.

| Major Meteor Showers | Peak Night | Time to Watch | Max. Rate* |
|---|---|---|---|
| | (may vary by a day) | (northern hemisphere) | (per hour) |
| Quadrantids | January 3-4 | 11:00 PM to dawn | 60-200 |
| Lyrids | April 21-22 | 9:30 PM to dawn | 10-15 typical |
| Eta Aquarids | May 5-6 | 1:30 AM to dawn | 40-85 |
| Southern Delta Aquarids | July 27-28 | 9:30 PM to dawn | 15-20 |
| Perseids | August 11-12 | dusk to dawn | 60-100 |
| Draconids/Giacobinids | October 7-8 | dusk to 2:00 AM or later | 10-20 typical |
| Orionids | October 20-21 | 10:00 PM to dawn | 25 |
| Leonids | November 17-18 | 11:30 PM to dawn | 10-15 |
| Geminids | December 13-14 | 7:00 PM to dawn | 60-120 |

* Under perfect conditions

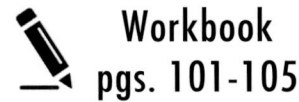

**Workbook pgs. 101-105**

A pair of interacting galaxies known as Arp 273. The larger galaxy is being distorted into a rose-like shape by the gravitational pull of the galaxy below it.

# Chapter 17

# Galaxies and the Universe

For all things were created at the beginning, being primordially woven into the texture of the world; but they await the proper opportunity for their existence.
— St. Augustine, *De Trinitate III.9*

## 17.1 The Milky Way

On a clear, dark night in the country, you can look up at the stars and see our **galaxy**, the **Milky Way**, spread across the sky. The Milky Way is a group of billions of stars, many of them surrounded by their own planets[1], asteroids, and comets. There are so many stars in the Milky Way that it is difficult to distinguish one star from another with the naked eye, just as it is difficult to distinguish one grain of sand from another from a distance. The many stars in our galaxy tend to blend into each other, forming a whitish streak across the sky that the ancient Greeks and Romans described as the "milky way."

A **galaxy** is a group of billions of stars.

The **Milky Way** is our home galaxy.

FIGURE 17.1 The Milky Way spread across the sky

Of course, the Milky Way is not actually made of milk, nor does it form a path across the sky. The Milky Way is shaped like a flying saucer with a thick center and long, pinwheel arms. The Milky Way is about 100,000 light-years across, which is the distance that light can travel in 100,000 years. It takes only one second for light to travel three times around the Earth, so 100,000 light-years is a tremendous distance.

---

[1] Although astronomers have looked, they have not found signs of life on any of these planets.

With a width of 100,000 light-years, the Milky Way is so huge that it can contain 150 billion stars without even being crowded. To better understand how many stars 150 billion really is, imagine a circular playground 50 feet across. In your imagination, cover the playground with sand to a depth of three feet. The number of grains of sand in this playground is the same as the number of stars in the Milky Way. That's a lot of stars!

Unlike the grains of sand in our imaginary playground, the stars in the Milky Way are quite far from each other. Imagine how large our imaginary playground would be if each grain of sand were three miles from every other grain of sand! That is how far apart the different stars in the Milky Way are, at least in the part of the galaxy where the Sun is. The average distance between the stars in our part of the Milky Way is about five light-years, which is the same as three miles to a grain of sand.

FIGURE 17.2  A spiral galaxy viewed edge-on

FIGURE 17.3 A chart of the Milky Way

## 17.2 More about the Milky Way

"Oooh, my brain hurts!" Mike groaned. "I'll never be able to grasp such huge distances and such big numbers!"

"No, you probably won't," Dad admitted. "I know I certainly haven't. We can still understand how a galaxy works, though.

"In some ways, the Milky Way is much like the solar system. In the solar system, everything revolves around the Sun, which is the center of the solar system. In the Milky Way, everything revolves around the bulge in the center of the galaxy. Since the Milky Way is so huge, it takes about 230 million years for a star to make a complete circle and return to where it started.

"Different galaxies have different shapes, but the Milky Way is a **spiral galaxy**. This is the name we give galaxies that have a central bulge surrounded by a thin disk. The stars in the disk are arranged in long 'arms' that make a spiral galaxy look like a giant pinwheel."

**Spiral galaxies** have a central bulge, surrounded by a thin disk with long, spiral arms.

"Where are we in the Milky Way?" Mike asked.

"We're near the edge, in one of the spiral arms," Dad said. "The Sun is located in the Perseus Arm of the Milky Way, on what is called the Orion Spur. This is about 28,000 light-years from the center."

"Too bad," Mike commented. "It'd be fun to be in the center of the galaxy."

"Fun is exactly what it would *not* be!" Dad exclaimed. "The stars are much closer together in the central bulge of the Milky Way than they are in the spiral arms. The first thing we'd notice if our solar system were suddenly moved to the **galactic** bulge is that it wouldn't be dark at night. We would be surrounded by too many bright stars!

"Of course, we could probably get used to perpetual daytime, but I don't think we could get used to being blown up regularly," Dad continued. "Since there are so many stars in the galactic bulge, there are also a lot more supernovae. As you know, supernovae are some of the most powerful explosions in the universe, and they are quite capable of wiping out all life on Earth. If the Earth had been located in the center of the galaxy, the human race would have been killed off many times over by nearby supernovae. As it is, our solar system is placed on the edge of one of the distant arms, where few supernovae occur."

"But I thought supernovae were good for the Earth," Mike said. "In Chapter 15 I learned that almost all the elements on Earth were produced in supernovae explosions."

**Galactic** means "belonging to or having to do with a galaxy."

FIGURE 17.4 Supernova 1994D on the outskirts of disk galaxy NGC 4526

"You're right—life on Earth really depends on long-past supernovae, because many of the elements on Earth were fused in the high temperatures of supernovae. Copper, nitrogen, and sodium are good examples. Other elements, such as carbon, oxygen, and iron, were fused inside of stars, but these elements would have remained locked up inside the stars forever if they hadn't been released by supernovae explosions. Nevertheless, if these supernovae had occurred nearby instead of far away, they would have destroyed the Earth instead of contributing to its formation."

"So the Earth could only have formed out of the elements distributed by supernovae," Mike summarized, "but it's a good thing the supernovae themselves took place a long time ago and a long way off."

"Exactly," Dad agreed.

## 17.3 The Birth of a Star

Now that you have studied the life and death of stars and the importance of supernovae for the formation of the Earth, it is time to learn how stars and planets are born. One reason we have not discussed this before is that scientists are not completely sure how stars and planets are formed, although they have some pretty good guesses. Another reason is that the formation of stars and planets is really an activity of the galaxy as a whole, so the topic had to wait until you learned about galaxies.

In a certain sense, the main "job" of a galaxy is to oversee the birth, death, and rebirth of stars. Of course, a star can't really be reborn. Once the Sun has turned into a white dwarf star, it will never be reborn as a young Sun surrounded by new planets. But the material that is released by dying stars in planetary nebulae and supernovae can eventually be recycled into new stars. Here's how astronomers believe it happens.

The Milky Way galaxy is full of stars, but it is also full of clouds of gas and dust. These clouds are known as **molecular clouds**. Some of the dust and gas in these molecular clouds has been around since the beginning of the galaxy, but a lot of it comes from dying stars that redistribute their matter throughout the galaxy in planetary nebulae and supernovae.

Molecular clouds are the "star nurseries" in the galaxy, the places where new stars are born. You have learned that the Sun is made up mostly of hydrogen and helium gases. This is because it was formed from the material in a molecular cloud, and molecular clouds consist mainly of hydrogen and helium gases.

**Molecular clouds** (muh-LEK-yuh-ler clouds) are the clouds of dust and gas in which new stars are born.

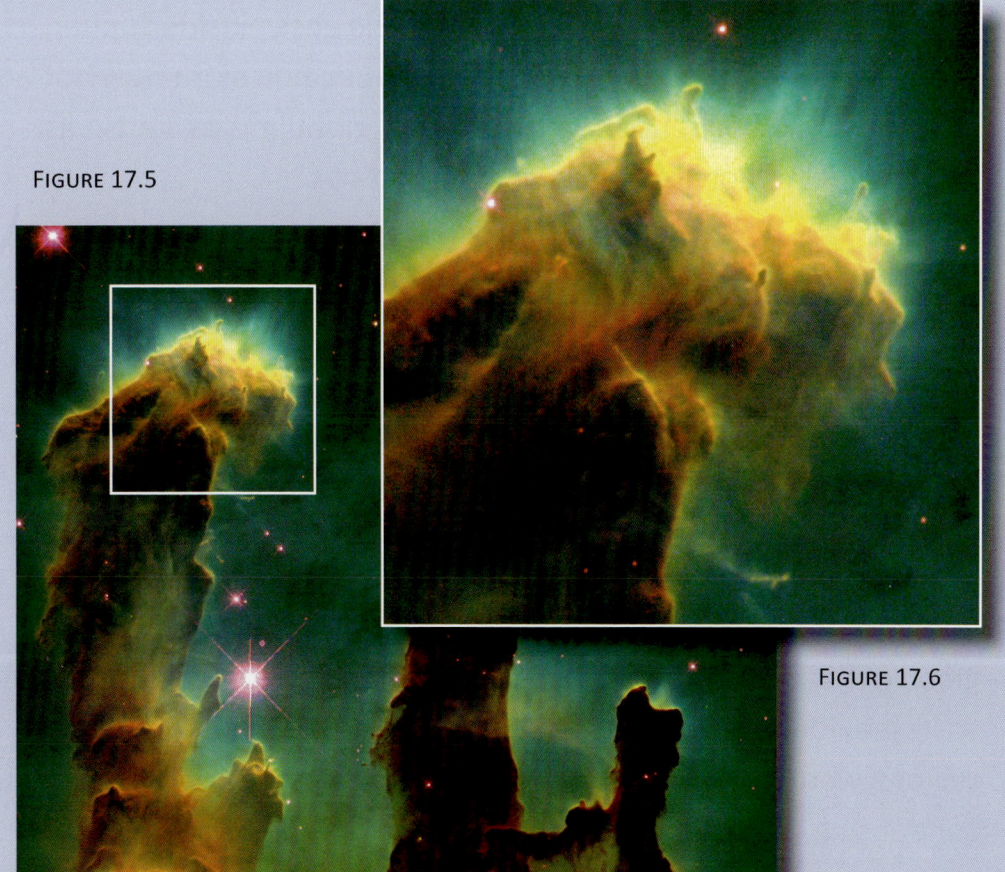

FIGURE 17.5

FIGURE 17.6

The golden brown "Pillars of Creation" in Figure 17.5 are part of a large molecular cloud known as the Eagle Nebula. The finger-like protrusions on the top of the left-most pillar (Figure 17.6) are globules in which new stars are forming. Each "fingertip" is larger than the entire solar system.

Figure 17.7 Star-forming molecular clouds

Now, you are probably wondering how a thin, drifting cloud of gas and dust can form a dense, burning star like our Sun. The answer is gravity. Nudged by shock waves from supernovae, the dust and gas in a molecular cloud start moving closer together. Since gravity is stronger when objects are closer together, the gravity of the dust and gas becomes stronger. Parts of the molecular cloud start to be pulled together by their own gravity into thick clumps of gas and dust called **globules**. As the globules contract, their gravity becomes even stronger and they begin to contract even more quickly, becoming denser and denser.

After several million years of being compressed by its own gravity, the center of a globule is as dense as the inside of Jupiter. As you remember from Chapter 16, the gases inside Jupiter are so dense that they act like liquids and metals instead of like gases. The gases inside of Jupiter are also very hot, because they are being crushed together by gravity. Like the gases inside Jupiter, the gases in the inner regions of a globule become extremely hot as the globule condenses. By the time a globule is as dense as Jupiter, it is well on its way to becoming a star.

Now, as you know, Jupiter is not a star, and it will never become a star. This is because Jupiter is not massive enough. It did not have enough gravity to condense farther once it got to the size it has now, so it never became hot enough to sustain nuclear fusion in its core.

Unlike Jupiter, the globule that formed our Sun *was* massive enough and *did* have enough gravity to become a star. In fact, the globule that became our Sun was 1000 times as massive as Jupiter, so its gravitational pull was extremely strong. Even after its inner regions were as dense as the inside of Jupiter, the globule kept condensing, and the farther it condensed, the hotter it became.

Eventually, the crushing gravity produced such high temperatures in the globule's core that nuclear fusion began. Our Sun was no longer a cloud of dust and gas. It was a newborn star, fusing hydrogen into helium in its core and shining by its own energy.

Meanwhile, some of the gas and dust on the outer edges of the globule did not end up as part of the Sun. Instead these "leftovers" became the eight planets, the asteroids, the comets, the moons, and all the rest of the solar system. The reason these objects travel in orbits around the Sun has to do with the motion of the original globule. Because of the way they form, globules usually end up rotating around their center as they condense. As a result, the planets and other objects that were once on the outer edges of our globule still travel in circles around the center of the solar system.

**Globules** (GLOB-yoolz) are the thick clumps of gas and dust that turn into new stars.

Workbook pgs. 107

Unit 3 Activity #8

## 17.4 Beyond the Milky Way

Within the Milky Way, star material is constantly recycled from new stars, to dying stars, and on to a new generation of stars. When a star dies, it releases much of its matter in a planetary nebula or a supernova. This matter then becomes part of a molecular cloud, and is formed into a new star inside a globule. This second star will eventually die, too, releasing its matter to be formed into more stars, and so on until the end of the universe.

FIGURE 17.8 The Hubble Deep Field Image is a view of a small part of the sky just above the Big Dipper.

The Milky Way is a very active place, but there is much more to the universe than just our own galaxy. The Milky Way is one among billions of galaxies. When astronomers look outside the Milky Way with special telescopes, they see a sky sprinkled with galaxies just as the sky on Earth is sprinkled with stars. The bright disks and specks in Figure 17.8 are the hundreds of

galaxies that can be seen in just a tiny part of the sky. Scientists were able to count 1500 galaxies in the Hubble Deep Field Image, and they estimate that there are 80 billion galaxies in the part of the universe that they can see from Earth. Each of these galaxies is located several million light-years from every other galaxy.

In Figure 17.8, the galaxies seem to be sprinkled at random across the sky, but this is not the case. Just as God ordered the geometric perfection of the spider's web, so He also arranged the galaxies. The Milky Way belongs to a group of galaxies called the Local Group, which is about 10 million light-years across. The Local

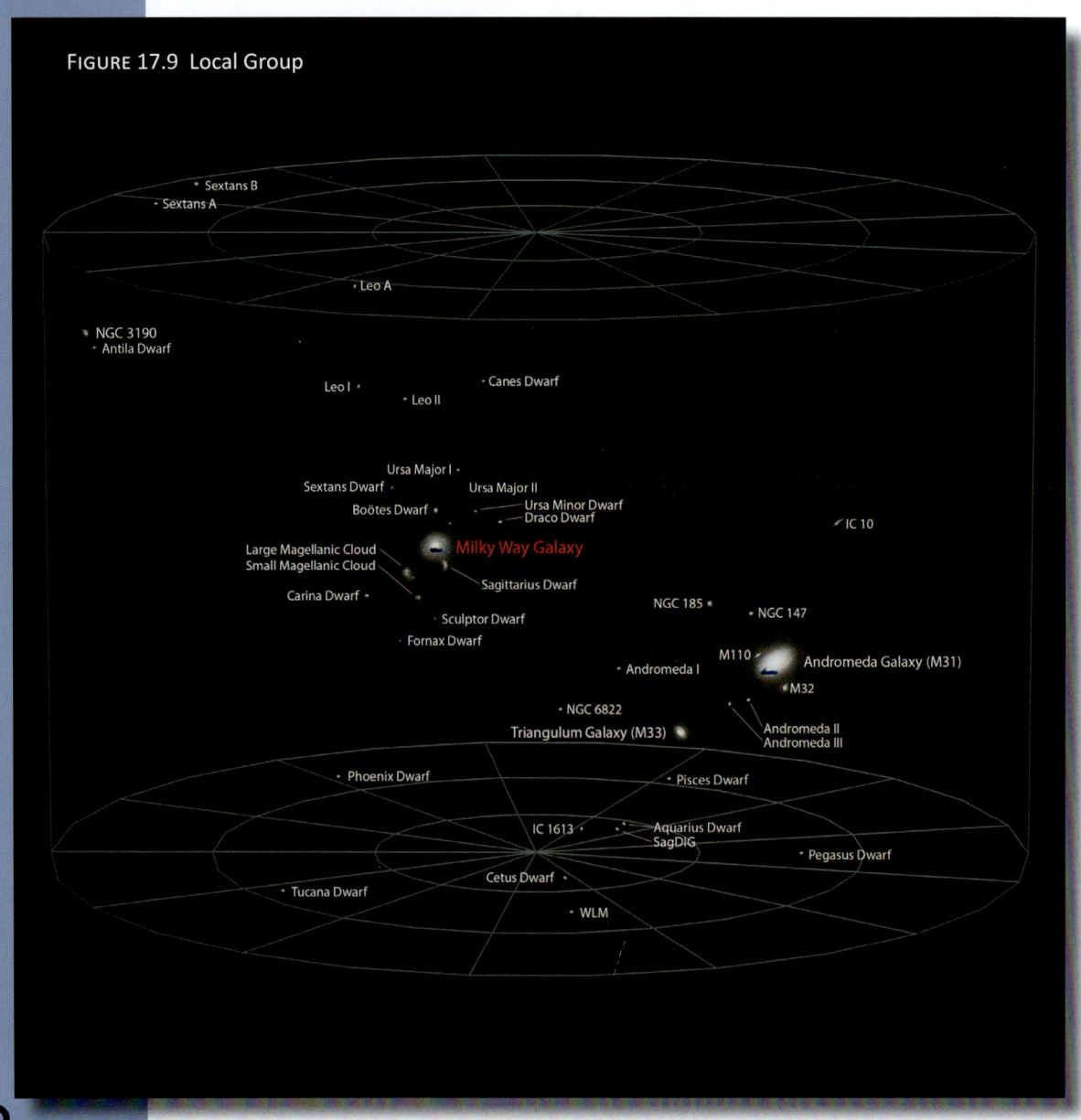

FIGURE 17.9 Local Group

Group contains about 40 galaxies, most of them smaller than the Milky Way (Figure 17.9).

The Local Group is part of a much larger structure called the Virgo Supercluster (Figure 17.10). The Virgo Supercluster contains millions of galaxies and groups of galaxies, each containing billions of stars. Still, the Virgo Supercluster is rather small compared to other superclusters. For instance, you can see in Figure 17.11 that the Ursa Major Supercluster, the Corona-Borealis Supercluster, and the Sextans Supercluster are all bigger than the Virgo Supercluster.

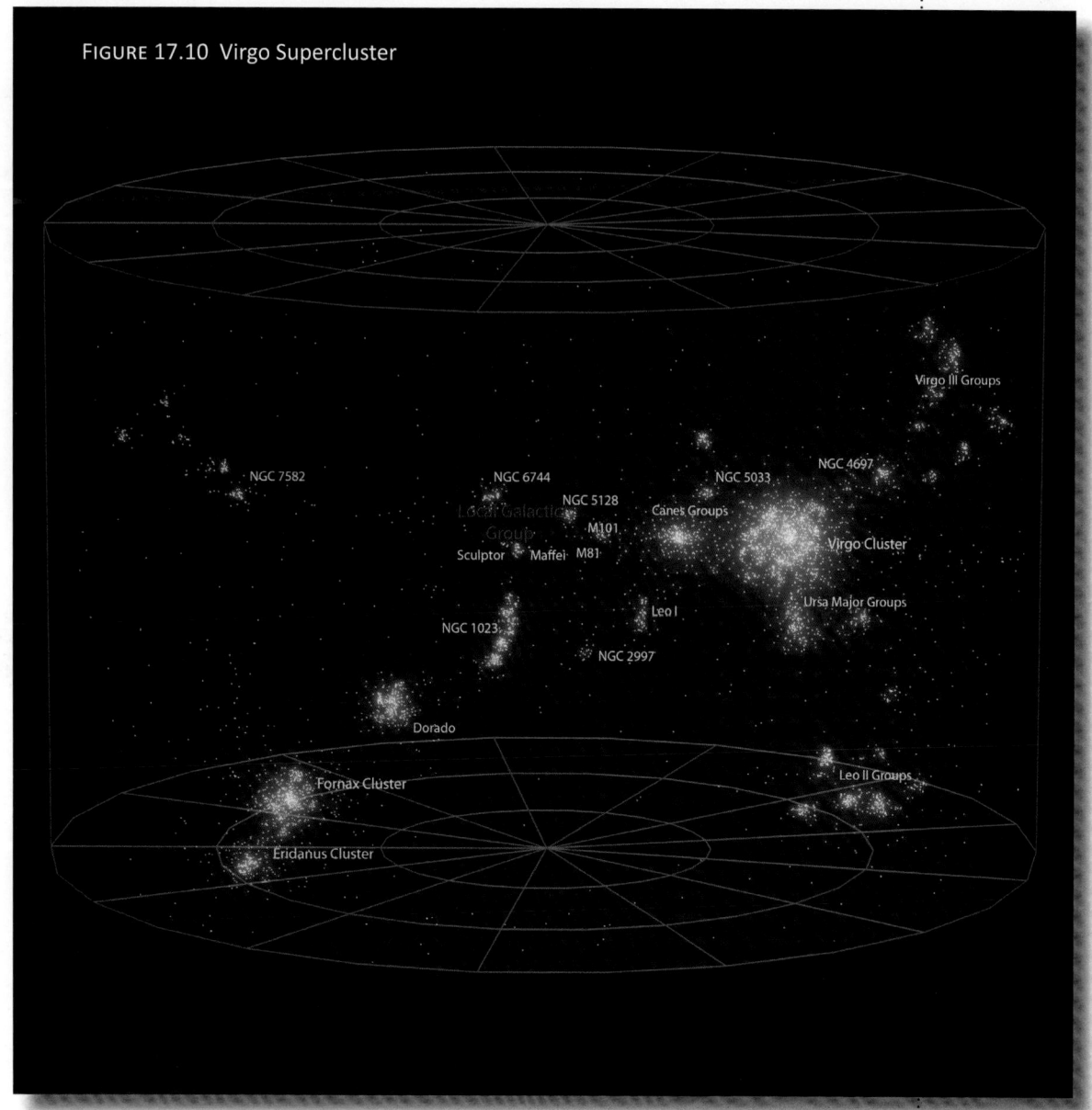

FIGURE 17.10 Virgo Supercluster

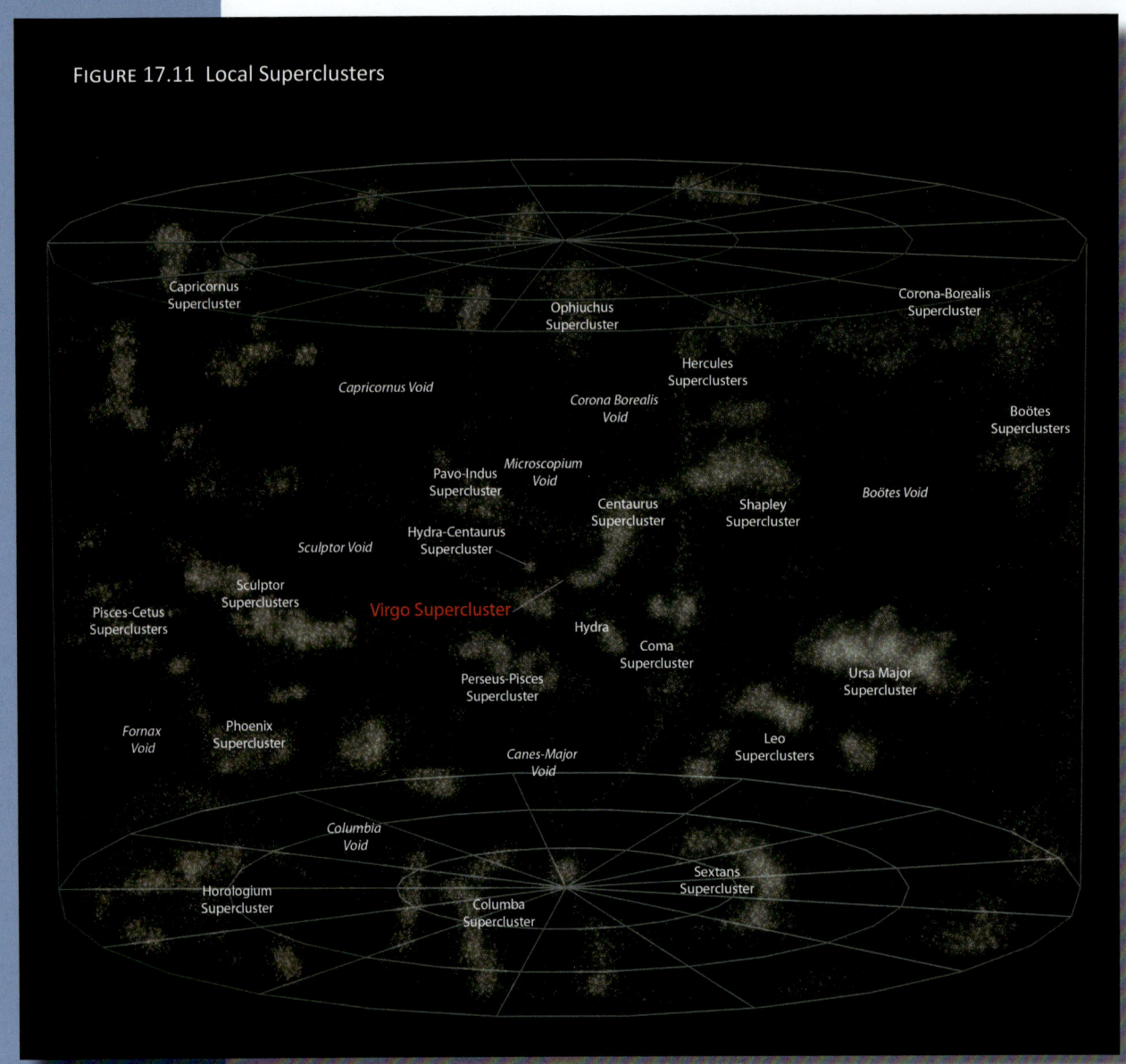

FIGURE 17.11 Local Superclusters

**A galactic wall** is a place where the superclusters are grouped together more closely than in other places. The groups of superclusters form patterns that resemble spider-web-like walls.

For a long time, astronomers thought that superclusters were the largest structures in the universe. In 1989, Margaret Geller discovered that the millions of superclusters are arranged in larger patterns called **galactic walls**, which surround huge voids, or empty spaces. (Several of these voids are labeled in Figure 17.11. Can you find the Sculptor Void, the Corona Borealis Void, and the Columbia Void?)

Astronomers do not know of any patterns larger than the galactic walls and the great voids, but new things are being discovered in astronomy every day! By the time you are grown up, scientists might have discovered that the universe is arranged in even larger and more impressive patterns. Still, we will never know everything about the universe. What we can know for certain is that God created the universe to be our home and that nothing in the universe—neither stars, galaxies, superclusters, nor even great walls—is as important as the soul of one human being.

## 17.5 The Beginning of the Universe

You may have heard of the **Big Bang theory**. This theory explains what the universe was probably like right after Creation, and how the universe came to be the way it is today.

According to the Big Bang theory, the universe began as a huge explosion, which was the beginning of space and time. At the beginning of the explosion, all the matter and energy in the present universe was contained as energy in the size of a single proton. (A proton, you remember, is 100,000 times smaller than an atom, and atoms are so small that a sheet of plastic wrap is about 100,000 atoms thick.) This proton-sized universe was so dense that scientists call it a **singularity**, which is a fancy word for a place that is so dense and small that scientists have no idea what goes on there.

Since the Big Bang was an explosion of all the matter and energy in the universe, it was much more powerful than all supernovae put together. The newborn universe was unbelievably hot and energetic, and it expanded with tremendous speed. In fact, the universe may have expanded from the size of a proton to the size of our solar system within a fraction of a second.

When the universe was first expanding, it consisted of an extremely hot "soup" of tiny particles of matter and energy. In fact, the protons, neutrons, and electrons in the early universe

**Workbook pg. 108**

**Unit 3 Activity #9**

The **Big Bang theory** says that the universe, including space and time, began as a huge explosion.

A **singularity** is a place that is so dense and small that scientists have no idea what goes on there.

**Big Bang Theory**

In everyday language, the word "theory" means an educated guess. In scientific language, "theory" means an hypothesis, or guess, that has been confirmed by experiments. The Big Bang theory is supported by a great deal of evidence, so it is much more than a guess. Of course, astronomers are still researching the details of the Big Bang, so the theory has been adjusted several times as new discoveries have been made. Nevertheless, astronomers are fairly certain that they are on the right track.

were too hot and energetic even to "stick together" as atoms. But as the universe expanded and became less dense, it also became cooler. Eventually, the universe became cool enough for protons, neutrons, and electrons to "stick together" into hydrogen and helium atoms, and these atoms formed huge clouds of hydrogen and helium.

During the next billion years, the universe continued to expand, and the clouds of hydrogen and helium gases became cool enough to condense into stars and galaxies. This happened in much the same way that stars form today inside molecular clouds (see section 17.3, The Birth of a Star). In fact, much of the gas in molecular clouds is a remnant of the hydrogen and helium clouds in the early universe.

Once galaxies were formed, the star-gas-star cycle began. New stars were born, grew old, and exploded as supernovae. The dust and gas from the supernovae were incorporated into molecular clouds, and then were reborn as a new generation of stars. Scientists currently estimate that the universe is 13.75 billion years old, so the cycle of star-gas-star has been going on in galaxies for nearly 13 billion years now.

The universe continued to expand once galaxies were formed, and it is still expanding today. All the galactic clusters in the universe are moving away from each other at tremendous speeds. However, this does not mean that all the stars in the universe are moving away from each other. The gravitational pull of each galaxy holds all its stars together in a single group. So even though the other galaxies are moving away from the Milky Way, the other stars in the Milky Way are not moving away from the Sun.

## 17.6 Evidence and Implications

You might wonder how astronomers can know what happened at the beginning of the universe. After all, they weren't there when it happened. What evidence is there that the universe began as the expansion of a singularity?

In the early part of the twentieth century, several pieces of evidence made astronomers wonder if all the clusters of galaxies in the universe might be moving away from each other. If this were true, it would mean that the universe is expanding, and scientists would need to figure out why.

In 1929, Edwin Hubble provided definite proof that the universe is expanding. Luckily, one scientist had already discovered an explanation for how and why this might happen. In 1927, Georges Lemaitre, a Catholic priest and a brilliant physicist, proposed the Big Bang theory as an explanation for the expansion of the universe. His theory said the universe was expanding because all the matter and energy in the universe had begun as a tiny singularity.

What does the expansion of the universe have to do with the Big Bang? Think about it. If we "rewind" the expansion of the universe in our imaginations, we come to a time when everything in the universe was crushed together in a single, tiny "speck," or singularity, of matter and energy. The moment when the singularity began to expand is the Big Bang. Our imaginations cannot "rewind" the universe any further than this point, because nothing was happening before the Big Bang, at least nothing

---

**How can scientists know the age of the universe?**

Scientists estimate the age of the universe by calculating how long it must have taken the universe to expand to its current state. Astronomers can also "date" the light that comes from stars, and some of the light they have seen is 13 billion years old. So they know the universe cannot be younger than that.

At first glance, the discovery that the world is billions of years old may seem to contradict what Genesis says about the world being created in six days. Since the Bible was not written as a science book, the Church teaches that we do not have to believe that the universe was created in six, 24-hour days. God is outside of time, so we do not know how long a "day" is to Him. As Scripture says, "with the Lord one day is like a thousand years and a thousand years like one day" (2 Peter 3:8).

## Church Teaching on Creation

The world and all things which are contained in it, both spiritual and material, were produced, according to their whole substance, out of nothing by God.

— Vatican I, *Canons on God the Creator of All Things*

We believe that God created the world according to his wisdom. It is not the product of any necessity whatever, nor of blind fate or chance.

— *Catechism of the Catholic Church*, 295

that we can know and measure through science. It was a "day without a yesterday," the beginning of the universe.

Georges Lemaitre proposed his theory of the "day without a yesterday" to explain why the universe was expanding, but many of Lemaitre's fellow scientists greatly disliked his theory. In fact, the name "Big Bang" is actually a phrase that one scientist used to make fun of Lemaitre's theory.

Probably the main reason Lemaitre's theory was so unpopular is that most of the scientists of his time believed the world was eternal. In other words, they thought the universe had always existed and always would because it was self-creating. But as Lemaitre's theory received more and more confirmation from astronomical observations, most scientists eventually accepted it. Most scientists now recognize the Big Bang theory as the best model they have to describe the beginning of the universe.

## 17.7 Creation and the Big Bang

Sometimes good Christians misunderstand the Big Bang theory and imagine that it is meant to show that God did not create the world. They would be surprised to learn that many scientists were prejudiced against the Big Bang theory precisely because it suggests that the world was created instead of always existing. As the *Catechism of the Catholic Church* explains, "the same God who reveals mysteries and infuses faith has bestowed the light of reason on the human mind," so we need never fear that the discoveries of science will contradict the truths of revelation (*CCC* 159). Rather, scientific discoveries about the origins of the world "invite us to even greater admiration for the greatness of the Creator, prompting us to give him thanks for all his works and for the understanding and wisdom he gives to scholars and researchers" (*CCC* 283).

> The question about the origins of the world and of man has been the object of many scientific studies which have splendidly enriched our knowledge of the age and dimensions of the cosmos, the development of life-forms and the appearance of man.
>
> — *Catechism of the Catholic Church*, 283

Of course, the Big Bang theory is not proof for Creation or for the existence of God. Science is a way of measuring and understanding physical space and time, so it is not its job to answer the question of whether anything or Anyone existed before space and time. Rather than being proof for Creation, the Big Bang theory is a description of what happened immediately after Creation—of what happened the moment after God said, "Let there be light."

Even though we cannot prove God's Creation of the universe through scientific experiments, reason tells us that the existence of the universe requires the existence of a First Cause. Everything that happens has to have a cause, and every cause (except the First Cause) also has to have a cause. The First Cause is the cause of all other causes. The First Cause did not need to be caused by something else, because the First Cause is eternal—it always was and always will be.

Through scientific observation we learn how the universe began, and through reason we learn of the existence of a First Cause. Finally and most importantly, we learn through faith that the First Cause loves each of us personally and cares for us as His children. As the *Catechism* explains, "The universe . . . is destined for and addressed to man, himself created in the 'image of God' and called to a personal relationship with God. . . . With creation, God does not abandon his creatures to themselves. He not only gives them being and existence, but also, and at every moment, upholds and sustains them in being, enables them to act and brings them to their final end" (*CCC* 299–301).

## 17.8  Our Place in the Universe

"Dad," Mike said thoughtfully, "when I think about the hugeness of the universe, I feel really unimportant and sort of, well, lost. Why does the universe have to be so big? What's the point of millions of galaxies and superclusters?"

"Well, for one thing, as man discovers more about the universe, it gives him a renewed sense of wonder and respect for God's infinite power," Dad smiled. "More importantly, our little planet Earth simply could never have existed if the universe weren't just the size it is."

"Really?" Mike asked, surprised.

"Really," Dad affirmed. "The Big Bang was not just a random explosion involving random amounts of energy and matter. Far from it. Every part of the Big Bang had to be very exact.

"If the universe were less massive—that is, if it had less 'stuff' in it—the Big Bang would have blown all the protons and atoms apart so quickly that there would be no time for galaxies, stars, and planets to form. You could compare this sort of universe to a balloon that you blow up so quickly and vigorously that it pops.

"On the other hand," Dad continued, "if the universe were more massive—that is, heavier, with more 'stuff' in it—the Big Bang would not have been powerful enough to make it expand. If the newborn universe expanded at all, it would quickly be pulled together again by its own gravity and collapse back into a dense clump. This 'heavy' universe would be like a very tight, thick balloon, that you blow up a little way and then give up on. After expanding to hold one or two breaths of air, the balloon collapses back into its original size."

"So the universe had to be as big as it is or the Earth would never have formed," Mike summarized.

"Precisely," Dad said. "And within our perfectly-sized universe, the Earth is positioned in just the right place. We orbit the Sun at just the right distance to stay warm, but not so close that we are scorched by the heat.

"Our Sun is just the right size, too. Stars that are larger than the Sun have so much mass that they are short-lived and often unstable, and stars that are smaller than the Sun give off much less light and heat. Since the Sun is an average-sized star, it is a reliable source of energy that gives out just the right amount of light and heat to provide for life on Earth.

"As you have learned, our solar system is located on the edge of one of the galactic arms. In this out-of-the-way corner, grass, puppies, and children can grow without fear of being annihilated by the supernovae in the center of the galaxy.

"We are also in just the right type of galaxy. If we were in a smaller galaxy, there might not have been enough supernovae to produce the elements needed for life on Earth. But a galaxy larger than the Milky Way would probably have too many supernovae for safety!

**The Big Bang explosion had to be exact to about 50 decimal places.**
That means that if the universe had 1/100,000,000,000,000,000,000,000,000,000,000,000,000,000,000,000 less matter, it would have been blown apart and stars and planets would never have formed. If the universe had 1/100,000,000,000,000,000,000,000,000,000,000,000,000,000,000,000 more matter, it would have collapsed back on itself.

"In other words, the huge size of the universe and our out-of-the-way location are actually signs of the special care God took in designing our home in the universe."

Mike jumped up and headed for the door. "I have to go tell that to Nick and Christie!" he exclaimed. "I can't wait until this summer's star-gazing trip when I get to teach them all about astronomy!"

*When I look at Your heavens, the work of Your fingers,
the moon and the stars which You have established;
What is man that You are mindful of him,
and the son of man that You care for him?
— Psalm 8:3-4*

**Workbook
pgs. 109-113**

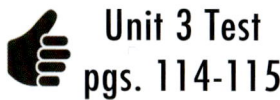

**Unit 3 Test
pgs. 114-115**

**Image Credits:**

Cover: © Johannes Kroemer / Tetra Images / Corbis; cover: © iStockphoto / Thinkstock; cover: © Hemera / Thinkstock; cover: NASA, ESA, J. Hester, A. Loll (ASU); cover: © Jupiterimages / Thinkstock; pg. i: © iStockphoto / Thinkstock; pg. i: © Hemera / Thinkstock; pg. i: NASA, ESA, J. Hester, A. Loll (ASU); pg. i: © Jupiterimages / Thinkstock; pg. iii: NASA, ESA, J. Clarke (Boston University), and Z. Levay (STScI); pg. iii: NASA, J. Bell (Cornell U.) and M. Wolff (SSI); pg. iii: NASA/JPL/University of Arizona; pg. iii: NSSDC Photo Gallery; pg. iii: The Visible Earth (http://visibleearth.nasa.gov/) Image by Reto Stockli, Alan Nelson, and Fritz Hasler; pg. 2-3: © ZERO CREATIVES / Photo Researchers, Inc.; © pg. 5: Hemera / Thinkstock; pg. 6-8: © Danny E Hooks / Shutterstock.com; pg. 9: © Dorling Kindersley RF / Thinkstock; pg. 10: © Hemera / Thinkstock; pg. 11: © Fidelis Schabet [Public domain], via Wikimedia Commons; pg. 11: © iStockphoto / Thinkstock; pg. 14: © Hemera / Thinkstock; pg. 14: © Hemera / Thinkstock; pg. 15: © iStockphoto / Thinkstock; pg. 15: © iStockphoto / Thinkstock; pg. 16: © Thinkstock Images / Comstock / Thinkstock; pg. 16: The Visible Earth (http://visibleearth.nasa.gov/) Image by Reto Stockli, Alan Nelson, and Fritz Hasler; pg. 16: © iStockphoto / Thinkstock; pg. 17: © Hemera / Thinkstock; pg. 17: Pasieka / Photo Researchers, Inc.; pg. 18: © Danny E Hooks / Shutterstock.com; pg. 18 (x17) Hemera / Thinkstock; pg. 18: © Pasieka / Photo Researchers, Inc.; pg. 18: © iStockphoto / Thinkstock; pg. 18: © TongRo Image Stock / Thinkstock; pg. 18: © iStockphoto / Thinkstock; pg. 20: © Hemera / Thinkstock; pg. 20: © iStockphoto / Thinkstock; pg. 21: © sarah2 / Shutterstock.com; pg. 21: © Mau Horng / Shutterstock.com; pg. 21: © Matt Baker / Shutterstock.com; pg. 22-24: © Charles D. Winters / Photo Researchers, Inc.; pg. 24: iStockphoto / Thinkstock; pg. 24: © iStockphoto / Thinkstock; pg. 24: © iStockphoto / Thinkstock; pg. 25: © John Foxx / Stockbyte / Thinkstock; pg. 27: © iStockphoto / Thinkstock; pg. 28: © iStockphoto / Thinkstock; pg. 28: © iStockphoto / Thinkstock; pg. 29: © Ingram Publishing / Thinkstock; pg. 29: © iStockphoto / Thinkstock; pg. 30: © Steve Lovegrove / Shutterstock.com; pg. 31: © iStockphoto / Thinkstock; pg. 31: © iStockphoto / Thinkstock; pg. 31: © iStockphoto / Thinkstock; pg. 31: © Charles D. Winters / Photo Researchers, Inc.; pg. 31: © Toth Tamas / Shutterstock.com; pg. 32: © iStockphoto / Thinkstock; pg. 32: © iStockphoto / Thinkstock; pg. 33: © iStockphoto / Thinkstock; pg. 34: © Danny E Hooks / Shutterstock.com; pg. 35: © Martyn F. Chillmaid / Photo Researchers, Inc.; pg. 36: © ITStock Free / Polka Dot / Thinkstock; pg. 37: © iStockphoto / Thinkstock; pg. 39: © iStockphoto / Thinkstock; pg. 40: © iStockphoto / Thinkstock; pg. 42-44: © Monkey Business Images / Shutterstock.com; pg. 44: © Creatas Images / Creatas / Thinkstock; pg. 45: © iStockphoto / Thinkstock; pg. 46: © Comstock / Comstock / Thinkstock; pg. 47: © iStockphoto / Thinkstock; pg. 47: © Brand X Pictures / Brand X Pictures / Thinkstock; pg. 48: © Hemera / Hemera / Thinkstock; pg. 49: © Evoken / Shutterstock.com; pg. 49: © iStockphoto / Thinkstock; pg. 49: © Joe Belanger / Shutterstock.com; pg. 50: © Blaj Gabriel / Shutterstock.com; pg. 50: © iStockphoto / Thinkstock; pg. 50: © iStockphoto / Thinkstock; pg. 51: © Pavel Shchegolev / Shutterstock.com; pg. 54: © Gildmai / Shutterstock.com; pg. 54: © Jupiterimages / Polka Dot / Thinkstock; pg. 56: © iStockphoto / Thinkstock; pg. 57: © Charles D. Winters / Photo Researchers, Inc.; pg. 58: © lyf1 / Shutterstock.com; pg. 59: © iStockphoto / Thinkstock; pg. 59: © iStockphoto / Thinkstock; pg. 60-62: © iStockphoto / Thinkstock; pg. 62: © iStockphoto / Thinkstock; pg. 63: © iStockphoto / Thinkstock; pg. 66: © iStockphoto / Thinkstock; pg. 67: © Jupiterimages / Comstock / Thinkstock; pg. 68: © iStockphoto / Thinkstock; pg. 69: © iofoto / Shutterstock.com; pg. 69: © Andrew Lambert Photography / Photo Researchers, Inc.; pg. 70: © Cordelia Molloy / Photo Researchers, Inc.; pg. 71: © worradirek / Shutterstock.com; pg. 72: © Cordelia Molloy / Photo Researchers, Inc.; pg. 72: © Cordelia Molloy / Photo Researchers, Inc.; pg. 72: © Cordelia Molloy / Photo Researchers, Inc.; pg. 73: © Ted Kinsman / Photo Researchers, Inc.; pg. 75: © Stockbyte / Stockbyte / Thinkstock; pg. 75: © Thinkstock Images / Comstock / Thinkstock; pg. 76: © Andrew McDonough / Shutterstock.com; pg. 78: © Doug Martin / Photo Researchers, Inc.; pg. 81: © Monty Rakusen / Photo Researchers, Inc.; pg. 81: © Stockbyte / Stockbyte / Thinkstock; pg. 82: © Hemera / Thinkstock; pg. 83: © iStockphoto / Thinkstock; pg. 83: © Hemera / Thinkstock; pg. 83: © iStockphoto / Thinkstock; pg. 83: © iStockphoto / Thinkstock; pg. 84: PD-Art, PD-1923 via Wikimedia Commons; pg. 84: © Photos.com / Photos.com / Thinkstock; pg. 85: © Photos.com / Photos.com / Thinkstock; pg. 85: By Henry Howard (1769 - 1847) [Public domain], via Wikimedia Commons; pg. 86-88: © Stockbyte / Stockbyte / Thinkstock; pg. 89: © Jupiterimages / Comstock / Thinkstock; pg. 90: © iStockphoto.com / Christina Richards; pg. 91: © Peter Daniello / ShutterPoint Stock Photography; pg. 92: © J.Gatherum / Shutterstock.com; pg. 92: © Horiyan / Shutterstock.com; pg. 93: Comstock / Comstock / Thinkstock; pg. 93: © tlorna / Shutterstock.com; pg. 93: Jupiterimages / Thinkstock; pg. 94: © Sergey Lavrentev / Shutterstock.com; pg. 95: © iStockphoto / Thinkstock; pg. 96: © Andriano / Shutterstock.com; pg. 97: © iStockphoto / Thinkstock; pg. 97: © Photos.com / Gettyimages / Thinkstock; pg. 97: © Goodshoot / Goodshoot / Thinkstock; pg. 98: © Stockbyte / Stockbyte / Thinkstock; pg. 98: © Hemera / Thinkstock; pg. 99: © Unlisted Images / Fotosearch.com; pg. 99: © Unlisted Images / Fotosearch.com; pg. 100: © Eyecandy Images / Thinkstock; pg. 100: © Stockbyte / Stockbyte / Thinkstock; pg. 100: © Comstock / Comstock / Thinkstock; pg. 100: © iStockphoto / Thinkstock; pg. 101: © Jiri Hera / Shutterstock.com; pg. 102: © Racheal Grazias / Shutterstock.com; pg. 103: © iStockphoto / Thinkstock; pg. 105: © iStockphoto / Thinkstock; pg. 106-107: © Vasily Kovalev / Shutterstock.com; pg. 108: © Wild Arctic Pictures / Shutterstock.com; pg. 109: © Steffen Foerster Photography / Shutterstock.com; pg. 110: © Darryl Brooks / Shutterstock.com; pg. 111: © iStockphoto / Thinkstock; pg. 111: © iStockphoto / Thinkstock; pg. 111: © Anup Shah / Digital Vision / Thinkstock; pg. 112: © iStockphoto / Thinkstock; pg. 112: © iStockphoto / Thinkstock; pg. 112: © Jupiterimages / Creatas / Thinkstock; pg. 113: © Hemera / Thinkstock; pg. 113: Tom Brakefield / Stockbyte / Thinkstock; pg. 113: © AbleStock.com / AbleStock.com / Thinkstock; pg. 114: © iStockphoto / Thinkstock; pg. 114: © Tom Brakefield / Stockbyte / Thinkstock; pg. 114: © iStockphoto / Thinkstock; pg. 114: © iStockphoto / Thinkstock; pg. 114: © Jupiterimages / Creatas / Thinkstock; pg. 115: © BananaStock / BananaStock / Thinkstock; pg. 115: © iStockphoto / Thinkstock; pg. 115: © iStockphoto / Thinkstock; pg. 115: © iStockphoto / Thinkstock; pg. 116: © Dorling Kindersley RF / Thinkstock; pg. 117: © Dorling Kindersley RF / Thinkstock; pg. 118: © Dorling Kindersley RF / Thinkstock; pg. 119: © Dorling Kindersley RF / Thinkstock; pg. 119: © Thomas Northcut / Digital Vision / Thinkstock; pg. 119: © Jupiterimages / Photos.com / Thinkstock; pg. 119: © iStockphoto / Thinkstock; pg. 119: © iStockphoto / Thinkstock; pg. 120-122: © Thomas Barrat / Shutterstock.com; pg. 123: © iStockphoto / Thinkstock; pg. 123: © Hemera Technologies / Photos.com / Thinkstock; pg. 124: © Dorling Kindersley RF / Thinkstock; pg. 125: © Dr. Juerg Alean / Photo Researchers, Inc.; pg. 126: © Wild Arctic Pictures / Shutterstock.com; pg. 126: image courtesy of Dr. Kenneth H. Dunton, Marine Science Institute, The University of Texas at Austin; pg. 127: © iStockphoto / Thinkstock; pg. 127: © iStockphoto / Thinkstock; pg. 127: © altrendo nature / Stockbyte / Thinkstock; pg. 128: © Wild Arctic Pictures / Shutterstock.com; pg. 128: © Nadezhda Bolotina / Shutterstock.com; pg. 128: © U.S. Fish and Wildlife Service / Alaska Region Library; pg. 129: © W. Lynch / ArcticPhoto; pg. 129: © Tom & Pat Leeson / Photo Researchers, Inc.; pg. 129: © Art Wolfe / Photo Researchers, Inc.; pg. 130: © Dorling Kindersley RF / Thinkstock; pg. 130: © iStockphoto / Thinkstock; pg. 130: © iStockphoto / Thinkstock; pg. 130: © iStockphoto / Thinkstock; pg. 131: © Arto Hakola / Shutterstock.com; pg. 132: © U.S. Fish and Wildlife Service / Matthew Perry; pg. 132: © Andrew Syred / Photo Researchers, Inc.; pg. 132: © iStockphoto / Thinkstock; pg. 133: © AbleStock.com / Photos.com / Thinkstock; pg. 134: © Selena / Shutterstock.com; pg. 134: © erashov / Shutterstock.com; pg. 135: © Andrew Syred / Photo Researchers, Inc.; pg. 136: © iStockphoto / Thinkstock; pg. 136: © Design Pics / Thinkstock; pg. 136: © iStockphoto / Thinkstock; pg. 136: © Wild Arctic Pictures / Shutterstock.com; pg. 137: © Hellen Sergeyeva / Shutterstock.com; pg. 137: © Kim Hansen (Own work) [GFDL (http://www.gnu.org/copyleft/fdl.html), CC-BY-SA-3.0 (creativecommons.org/licenses/by-sa/3.0/), via Wikimedia Commons; pg. 137: © Wild Arctic Pictures / Shutterstock.com; pg. 139: © Nickolay Stanev / Shutterstock.com; pg. 139: © Nana77777 / Shutterstock.com; pg. 139: © Dr U (Own work) [Public domain], via Wikimedia Commons; pg. 139: © 4028mdk09 (Own work) [CC-BY-SA-3.0 (creativecommons.org/licenses/by-sa/3.0)], via Wikimedia Commons; pg. 139: © Mathias Karlsson (Own work) [GFDL (http://www.gnu.org/copyleft/fdl.html), CC-BY-SA-3.0 (creativecommons.org/licenses/by-sa/3.0/) via Wikimedia Commons; pg. 140: © Todd Boland / Shutterstock.com; pg. 141: © Vladimir Melnik / Shutterstock.com; pg. 142: © Andreas Gradin / Shutterstock.com; pg. 142: © Vasilevich Aliaksandr / Shutterstock.com; pg. 143: © Taina Sohlman / Shutterstock.com; pg. 143: © kruemel / Shutterstock.com; pg. 143: © FormosanFish / Shutterstock.com; pg. 143: © Vilmos Varga / Shutterstock.com; pg. 143: © Gorilla / Shutterstock.com; pg. 144-146: © Michael Giannechini / Photo Researchers, Inc.; pg. 146: © peupleloup [CC-BY-SA-2.0 (creativecommons.org/licenses/by-sa/2.0)], via Wikimedia Commons; pg. 147: © Red Circle Images RM / www.fotosearch.com; pg. 148: © Design Pics / Thinkstock; pg. 149: © U.S. Fish and Wildlife Service / Innoko NWR-030; pg. 149: © U.S. Fish and Wildlife Services / Steve Hillebrand; pg. 150: © Steve Bower / Shutterstock.com; pg. 150: © iStockphoto / Thinkstock; pg. 150: © Hemera / Thinkstock; pg. 150: © Image Source / Thinkstock; pg. 151: Elaine R. Wilson, www.naturespicsonline.com [CC-BY-SA-3.0 (creativecommons.org/licenses/by-sa/3.0)], via Wikimedia Commons; pg. 151: © Critterbiz / Shutterstock.com; pg. 152: © Dorling Kindersley RF / Thinkstock; pg. 152 © iStockphoto.com/stanley45; pg. 152: © iStockphoto / Simply Creative Photography; pg. 153: USDA [Public domain], via Wikimedia Commons; pg. 154: © iStockphoto / kawisphoto; pg. 154: © Gail Johnson / Shutterstock.com; pg. 155: iStockphoto / DeniseBush; pg. 156: © oksix / Shutterstock.com; pg. 156: © Jupiterimages / Creatas / Thinkstock; pg. 157: © jokerpro / Shutterstock.com; pg. 157: © iStockphoto / Thinkstock; pg. 158: © jayfish / Shutterstock.com; pg. 159: © Hemera / Thinkstock; pg. 159: © iStockphoto / Thinkstock; pg. 160: USDA-NRCS PLANTS Database / Herman, D.E. et al. 1996. North Dakota tree handbook. USDA NRCS ND State Soil Conservation Committee; NDSU Extension and Western Area Power Admin., Bismarck, ND. ([1]) [Public domain], via Wikimedia Commons; pg. 161: © iStockphoto / Thinkstock; pg. 161: © Pi-Lens / Shutterstock.com; pg. 162: © Andrea Moro / CC BY-SA (https://creativecommons.org/licenses/by-sa/3.0); pg. 162 © Pascal Goetgheluck / ardea.com; pg. 162: © Hemera / Thinkstock; pg. 163: © Ariel Bravy / Shutterstock.com; pg. 163: © iStockphoto / Thinkstock; pg. 163: © scaners3d / Shutterstock.com; pg. 164: © iStockphoto / Thinkstock; pg. 165: Image Courtesy Green Water Fishing Adventures greenwaterguides.com; pg. 166: © iStockphoto / Thinkstock; pg. 166: © iStockphoto / Thinkstock; pg. 166: © iStockphoto / Thinkstock; pg. 168-170: © iStockphoto / Thinkstock; pg. 170: © Hemera / Thinkstock; pg. 171: © Hemera / Thinkstock; pg. 171: © iStockphoto / Thinkstock; pg. 171: © iStockphoto / Thinkstock; pg. 171: © iStockphoto / Thinkstock; pg. 172: © Karel Gallas / Shutterstock.com; pg. 172: © stockerman / Shutterstock.com; pg. 173: © altrendo nature / Stockbyte / Thinkstock; pg. 173: © KUCU / Shutterstock.com; pg. 174: © Holger Ehlers / Shutterstock.com; pg. 174: © Mike Rogal / Shutterstock.com; pg. 174: © Steve Byland / Shutterstock.com; pg. 175: © iStockphoto / Thinkstock; pg. 175: © jordache / Shutterstock.com; pg. 176: © Medioimages/Photodisc / Photodisc / Thinkstock; pg. 176: © iStockphoto / Thinkstock; pg. 176: © iliuta goean / Shutterstock.com; pg. 176: © iStockphoto / Thinkstock; pg. 176: Crested Green Basilisk © Emily S. Damstra; pg. 177: © Dr. Morley Read / Shutterstock.com; pg. 177:

© iStockphoto / Thinkstock; pg. 178: © Photos.com / Photos.com / Thinkstock; pg. 178: © iStockphoto / Thinkstock; pg. 179: © iStockphoto / Thinkstock; pg. 179: © iofoto / Shutterstock.com; pg. 180: © iStockphoto / Thinkstock; pg. 181: © Eky Studio / Shutterstock.com; pg. 181: © iStockphoto / Thinkstock; pg. 181: © Hemera Technologies / PhotoObjects.net / Thinkstock; pg. 182: © Anna Kucherova / Shutterstock.com; pg. 182: © Ammit / Shutterstock.com; pg. 182: © Dr. Morley Read / Shutterstock.com; pg. 182: © ngarare / Shutterstock.com; pg. 183: © Tom Brakefield / Stockbyte / Thinkstock; pg. 183: © iStockphoto / Thinkstock; pg. 183: © Hemera / Thinkstock; pg. 183: © Hemera / Thinkstock; pg. 183: © Photos.com / Thinkstock; pg. 184: © Anup Shah / Thinkstock; pg. 184: © iStockphoto / Brasil2; pg. 185: © Nina B / Shutterstock.com; pg. 185: © Joseph C Boone (Own work) [CC BY-SA 3.0 (http://creativecommons.org/licenses/by-sa/3.0)], via Wikimedia Commons; pg. 185: © Eric Gropp / CC-BY-SA-2.0 (https://creativecommons.org/licenses/by/2.0/); pg. 185: © Jupiterimages / Photos.com / Thinkstock; pg. 186: © iStockphoto / Thinkstock; pg. 186: © iStockphoto / NNehring; pg. 186: © Dr. Morley Read / Shutterstock.com; pg. 186: © Abhindia / Shutterstock.com; pg. 186: © iStockphoto / T-Immagini; pg. 186: © Andy Magee / Shutterstock.com; pg. 186: Andrew Gruswitz at en.wikipedia [CC-BY-SA-3.0 (creativecommons.org/licenses/by-sa/3.0) or GFDL (www.gnu.org/copyleft/fdl.html)], from Wikimedia Commons; pg. 186: © naturespicsonline.com; pg. 187: © Alex Wild/Visuals Unlimited/Corbis; pg. 188: © Seth Laster / Shutterstock.com; pg. 188: © Brandon Alms / Shutterstock.com; pg. 189: © iStockphoto / GraffiZone; pg. 190-192: © Craig K. Lorenz / Photo Researchers, Inc.; pg. 193: © iStockphoto / Thinkstock; pg. 193: © iStockphoto / Thinkstock; pg. 193: © Jupiterimages / Comstock / Thinkstock; pg. 194: © George Burba / Shutterstock.com; pg. 194: © Alfie Photography / Shutterstock.com; pg. 194: © Guy J. Sagi / Shutterstock.com; pg. 194: © iStockphoto / Thinkstock; pg. 195: © whitehoune / Shutterstock.com; pg. 196: © iStockphoto / Thinkstock; pg. 196: © Kitch Bain / Shutterstock.com; pg. 197: © Piotr Gatlik / Shutterstock.com; pg. 197: © iStockphoto / Thinkstock; pg. 197: © Hemera / Thinkstock; pg. 198: © Joy Stein / Shutterstock.com; pg. 198: © Jupiterimages / Photos.com / Thinkstock; pg. 198: © Hector Garcia Serrano / Shutterstock.com; pg. 199: © iStockphoto / Thinkstock; pg. 199: © Dorling Kindersley RF / Thinkstock; pg. 200: © Dorling Kindersley RF / Thinkstock; pg. 200: © Photodisc / Photodisc / Thinkstock; pg. 201: © Hemera / Thinkstock; pg. 202: © iStockphoto / Visual Communications; pg. 202: © iStockphoto / Dr-Strangelove; pg. 202: © iStockphoto / jupiterimages; pg. 202: Zoonar / Thinkstock; pg. 203: © bierchen / Shutterstock.com; pg. 203: © creativex / Shutterstock.com; pg. 204: © iStockphoto / Thinkstock; pg. 204: © Николай Усик / http://paradoxusik.livejournal.com / CC BY-SA (https://creativecommons.org/licenses/by-sa/3.0); pg. 204: © iStockphoto / Rainbohm; pg. 205: © Design Pics / Thinkstock; pg. 205: © kavram / Shutterstock.com; pg. 206: © naipung / Shutterstock.com; pg. 206: © U.S. Fish and Wildlife Service / Steve Hillebrand; pg. 207: Hans Hillewaert (Own work) [CC-BY-SA-3.0 (creativecommons.org/licenses/by-sa/3.0)], via Wikimedia Commons; pg. 207: © Arno Dietz / Shutterstock.com; pg. 207: © Francois Loubser / Shutterstock.com; pg. 207: © Chantelle Bosch/Shutterstock.com; pg. 208: © Michael & Patricia Fogden/CORBIS; pg. 208: © Honjune / Shutterstock.com; pg. 209: © Lisa M Smith / Shutterstock.com; pg. 210: © Jens Peermann / Shutterstock.com; pg. 211: © iStockphoto / Thinkstock; pg. 211: Luis Fernández García L. Fdez (Own work) [CC-BY-SA-2.1-es (creativecommons.org/licenses/by-sa/2.1/es/deed.en)], via Wikimedia Commons; pg. 211: © Vysokova Ekaterina / Shutterstock.com; pg. 212: Neal Parish from Oakland, CA (IMG_0794) [CC-BY-SA-2.0 (creativecommons.org/licenses/by-sa/2.0)], via Wikimedia Commons; pg. 213: © John A. Anderson / Shutterstock.com; pg. 213: © Antoni Murcia / Shutterstock.com; pg. 214: © AoiK / Shutterstock.com; pg. 215: © iStockphoto / Paul Tessier; pg. 216-218: © Photodisc / Photodisc / Thinkstock; pg. 218: Fanghong (Own work) [GFDL (http://www.gnu.org/copyleft/fdl.html), CC-BY-SA-3.0 (creativecommons.org/licenses/by-sa/3.0/) via Wikimedia Commons; pg. 219: © George Doyle / Stockbyte / Thinkstock; pg. 220: © iStockphoto / Thinkstock; pg. 221: © Ron Chapple Studios / Thinkstock; pg. 221: © Ron Chapple Studios / Thinkstock; pg. 222: © iStockphoto / Thinkstock; pg. 222: © iStockphoto / Thinkstock; pg. 222: © iStockphoto / Thinkstock; pg. 223: © iStockphoto / Thinkstock; pg. 224: © iStockphoto / Thinkstock; pg. 225: © Ingram Publishing / Thinkstock; pg. 226: © iStockphoto / Thinkstock; pg. 227: © Ingram Publishing / Thinkstock; pg. 227: © Hemera / Thinkstock; pg. 228: © iStockphoto / Thinkstock; pg. 228: International Rice Research Institute (IRRI) / CC BY (https://creativecommons.org/licenses/by/2.0); pg. 229: © iStockphoto / Thinkstock; pg. 231: © Hemera / Thinkstock; pg. 232-234: © iStockphoto / Thinkstock; pg. 234-235: © Hemera / Thinkstock; pg. 236: NASAImages.org; pg. 237: © Molodec / Shutterstock.com; pg. 238: © Nikm / Shutterstock.com; pg. 238: John R. Foster / Photo Researchers, Inc.; pg. 239: © Dorling Kindersley RF / Thinkstock; pg. 241: © Dorling Kindersley RF / Thinkstock; pg. 241: © iStockphoto / Thinkstock; pg. 241: © Viktar Malyshchyts / Shutterstock.com; pg. 242: © Markus Gann / Shutterstock.com; pg. 245: © Le Do / Shutterstock.com; pg. 245: © MO:SES; pg. 245: © Sue McDonald / Shutterstock.com; pg. 246: © Konstantin Mironov / Shutterstock.com; pg. 247: © Dorling Kindersley RF / Thinkstock; pg. 247: © Alexey Fursov / Shutterstock.com; pg. 248: © Dorling Kindersley RF / Thinkstock; pg. 249: © Hemera / Thinkstock; pg. 250: © iStockphoto / Thinkstock; pg. 251: © Dorling Kindersley RF / Thinkstock; pg. 252: © Dorling Kindersley RF / Thinkstock; pg. 253: © Dorling Kindersley RF / Thinkstock; pg. 253: Iztok Bončina/ESO (http://www.eso.org/public/images/potw1043a/) [CC-BY-3.0 (creativecommons.org/licenses/by/3.0) via Wikimedia Commons; pg. 254: LCGS Russ (Own work) [CC-BY-SA-3.0 (creativecommons.org/licenses/by-sa/3.0)], via Wikimedia Commons; pg. 254: © Dorling Kindersley RF / Thinkstock; pg. 256: © iStockphoto / Thinkstock; pg. 258: © Stocktrek Images / Thinkstock; pg. 260-262: NASA Langley Research Center (NASA-LaRC); pg. 263: NASA/ http://spaceflight.nasa.gov; pg. 263: Apollo 17 Crew, NASA; pg. 264: NASA/ http://spaceflight.nasa.gov; pg. 264: NASA Human Spaceflight Collection; pg. 264: NASA/ http://spaceflight.nasa.gov; pg. 265: © Dorling Kindersley RF / Thinkstock; pg. 267: NASA; pg. 268: NASA; pg. 268: © Dorling Kindersley RF / Thinkstock; pg. 269: By NASA/JSC [Public Domain], via Wikimedia Commons; pg. 270: © Taipan Kid / Shutterstock.com; pg. 274: © Dr Ajay Kumar Singh / Shutterstock.com; pg. 275: © Dr Ajay Kumar Singh / Shutterstock.com; pg. 275: © Ingrid Prats / Shutterstock.com; pg. 277: Manoj.dayyala (Own work) [CC-BY-SA-3.0 (creativecommons.org/licenses/by-sa/3.0) via Wikimedia Commons; pg. 278: NASA, via Wikimedia Commons; pg. 279: © Dorling Kindersley RF / Thinkstock; pg. 281: © godrick / Shutterstock.com; pg. 279: pg. 279: ph133 (Own work) [GFDL (www.gnu.org/copyleft/fdl.html), via Wikimedia Commons; pg. 279: Manoj.dayyala (Own work) [CC-BY-SA-3.0 (creativecommons.org/licenses/by-sa/3.0), via Wikimedia Commons; pg. 280: A013231 (Own work) [CC-BY-SA-3.0 (creativecommons.org/licenses/by-sa/3.0), via Wikimedia Commons; pg. 280: NASA/ http://spaceflight.nasa.gov; pg. 281: © Dorling Kindersley RF / Thinkstock; pg. 281: © godrick / Shutterstock.com; pg. 283: © iStockphoto / Thinkstock; pg. 283: © iStockphoto / Thinkstock; pg. 284-286: Evercat at en.wikipedia [Public domain], from Wikimedia Commons; pg. 286: NASA Human Spaceflight Collection; pg. 287: NASA Human Spaceflight Collection; pg. 289: © Snowbelle / Shutterstock.com; pg. 289: © Andrea Danti / Shutterstock.com; pg. 290: SOHO (ESA & NASA) / NASA Solar and Heliospheric Observatory Collection; pg. 290: © corepics / Shutterstock.com; pg. 291: © Pi-Lens / Shutterstock.com; pg. 292: NASA / Human Space Flight; pg. 294: Lsmpascal (Own work) [CC-BY-SA-3.0 (creativecommons.org/licenses/by-sa/3.0)], via Wikimedia Commons; pg. 294: © Mopic / Shutterstock.com; pg. 295: © iStockphoto / Thinkstock; pg. 297: SOHO (ESA & NASA) / NASA; pg. 298: © iStockphoto / Thinkstock; pg. 300: NASA, NOAO, ESA, the Hubble Helix Nebula Team, M. Meixner (STScI), and T.A. Rector (NRAO); pg. 301: NASA/The Hubble Heritage Team (STScI/AURA/NASA); pg. 301: Zoonar / Thinkstock; pg. 301: NASA and H. Richer (University of British Columbia); pg. 301: © Ron Chapple Studios / Thinkstock; pg. 304: NASA, ESA, J. Hester, A. Loll (ASU); pg. 305: Apollo 8 crewmember Bill Anders (NASA [1]), via Wikimedia Commons; pg. 306-308: NASA/ JPL/Space Science Institute; pg. 311: Lsmpascal (Own work) [CC-BY-SA-3.0 (creativecommons.org/licenses/by-sa/3.0)], via Wikimedia Commons; pg. 313: NASA/Johns Hopkins University Applied Physics Laboratory/Carnegie Institution of Washington; pg. 314: R. Stockli, A. Nelson, F. Hasler, NASA/GSFC/NOAA/USGS; pg. 316: NASA/JPL-Caltech; pg. 317: U.S. Navy photo by Mass Communication Specialist Seaman Chelsea Kennedy [Public domain], via Wikimedia Commons; pg. 319: NASA, J. Bell (Cornell U.) and M. Wolff (SSI); pg. 319: Stocktrek Images / Thinkstock; pg. 320: NASA/JPL/University of Arizona; pg. 321: By NASA/JPL-Caltech/JAXA/ESA, via Wikimedia Commons; pg. 321: Lunar and Planetary Institute; pg. 322: NASA/JPL-Caltech/UCLA/MPS/DLR/IDA; pg. 322: USGS Copyright Free Policy; pg. 323: NASA/JPL/Space Science Institute; pg. 324: NASA; pg. 325: Galileo Project, JPL, NASA; pg. 326: NASA/JPL/Space Science Institute; pg. 326: NASA/JPL-Caltech; pg. 327: NASA; pg. 328: NASA/JHUAPL/SWRI; pg. 329: NASA; pg. 330: Wally Pacholka / AstroPics.com; pg. 332-334: NASA, ESA, and the Hubble Heritage Team (STScI/AURA); pg. 334: © EpicStockMedia / Shutterstock.com; pg. 335: Bruce Hugo and Leslie Gaul, Adam Block (KPNO Visitor Program [www.noao.edu]), NOAO [http://www.noao.edu/], AURA, NSF [www.nsf.gov/]; pg. 336: NASA/JPL-Caltech; pg. 337: NASA, ESA, The Hubble Key Project Team, and The High-Z Supernova Search Team; pg. 339: NASA, ESA, STScI, J. Hester and P. Scowen (Arizona State University); pg. 340: ESO/Igor Chekalin; pg. 340: ESO/IDA/Danish 1.5 m/R. Gendler, J.-E. Ovaldsen, C. Thöne and C. Féron; pg. 340: ESO/Sergey Stepanenko; pg. 340: NASA, H. Ford (JHU), G. Illingworth (UCSC/LO), M.Clampin (STScI), G. Hartig (STScI), the ACS Science Team, and ESA; pg. 341: NASA, R. Williams and The Hubble Deep Field Team (STScI); pg. 341: © ella1977 / Shutterstock.com; pg. 342: Andrew Z. Colvin (Own work) [CC-BY-SA-3.0 (creativecommons.org/licenses/by-sa/3.0), via Wikimedia Commons; pg. 343: Andrew Z. Colvin (Own work) [CC-BY-SA-3.0 (creativecommons.org/licenses/by-sa/3.0), via Wikimedia Commons; pg. 344: Andrew Z. Colvin (Own work) [CC-BY-SA-3.0 (creativecommons.org/licenses/by-sa/3.0), via Wikimedia Commons; pg. 348: NASA, ESA, and the Hubble Heritage (STScI/AURA)-ESA/Hubble Collaboration; pg. 350: Stockbyte / Thinkstock; pg. 353: iStockphoto / Thinkstock; back cover: © oksana.perkins / Shutterstock.com; back cover: © ephotographer/Shutterstock; back cover: © oorka / Shutterstock.com; back cover: © Poznukhov Yuriy / Shutterstock.com; back cover: © markrhiggins / Shutterstock.com; back cover: © Valerie Potapova / Shutterstock.com